# Scientific Research

## as a

# Career

# Scientific Research
## as a
# Career

Finlay MacRitchie

**CRC Press**
Taylor & Francis Group
Boca Raton London New York

CRC Press is an imprint of the
Taylor & Francis Group, an **informa** business

CRC Press
Taylor & Francis Group
6000 Broken Sound Parkway NW, Suite 300
Boca Raton, FL 33487-2742

Printed in the United States of America on acid-free paper
10 9 8 7 6 5 4 3 2 1

International Standard Book Number: 978-1-4398-6965-9 (Paperback)

**Visit the Taylor & Francis Web site at**
**http://www.taylorandfrancis.com**

**and the CRC Press Web site at**
**http://www.crcpress.com**

# Contents

# Foreword

*Non plus ultra* ("nothing more beyond") was the warning to those who sailed the Mediterranean Sea. According to Greek mythology, these words were written on the Pillars of Hercules, at the Strait of Gibraltar, marking the edge of the then-known world.

The attitude was: "Go no further!" "Nothing more to discover!" Today, we might say: "We have hit a brick wall!"

Ignoring the ancient warning, the motto *Plus ultra* ("there *is* more beyond") was adopted by Charles V, King of Spain, following the discovery of the Americas.

But is there "more beyond" today? There are no new continents to discover, no new frontiers to colonize, no new chemical elements to be named. We have walked on the moon. We have probed the Solar System and beyond. Is exploration dead?

Have we now exhausted the possibilities for adventure, for discovery, for excitement, for exploration? Have we hit the proverbial "brick wall?" Is there really *Non plus ultra*?

No, no, no! Exploration is not dead. There is much "more beyond."

Arm yourself with the motto of Charles V: *Plus ultra*. But instead of voyaging beyond the limits of the Mediterranean, voyage beyond the limits of our scientific knowledge. Instead of donning the seaman's *sou'wester*, assume the mantle of the lab coat.

It is still possible to achieve the thrill of being able to say: "Sometimes I get to feel that I'm the only person in the world who fully understands a particular question." (See Chapter 1.)

The career in science can still lead us to the point of being able to say: "I am the first ever to demonstrate *XXX*." "I am the first ever to elucidate how *XXX* happens." What type of *XXX* do you choose?

Where would you wish the new exploration to lead you?

Just read on!

**Colin Wrigley, AM**

# Preface

This book is intended to present a picture of what is involved in a career in scientific research. The targeted readers are those who are already engaged in as well as those who are contemplating a career in science. It is also hoped that this book will attract readers from the general public and those who influence public policy. Some sections of the book are provocative, and readers will no doubt disagree with many of the opinions expressed. This is the nature of science. Contrary opinions are welcome, and new and valuable insights can often arise from the ensuing debate. This is in great contrast to some other disciplines. In science, precedence counts for nothing. Every theory and interpretation is continually questioned and challenged, regardless of when it was originally put forward. Scientific understanding is a dynamic process and always subject to change. The aim of research is to discover the truth and advance genuine knowledge. Scientific research enriches our culture and drives the technology that has improved the living conditions in which many humans live. It is therefore one of the most noble professions. Science is universal. The pursuit of knowledge should not be affected by national boundaries. This is one of the great attractions of working as a researcher. It enables formation of friendships with colleagues in various parts of the world and provides opportunities for international travel and for the nurturing of these friendships.

The greatest advances have been made by scientists when the conditions have allowed their creativity to flourish. Scientific research is an exploration into the unknown. This is not properly understood by many of those who are in charge. As a result, in recent decades, there has been an increasing encroachment of business management principles into the control of research. One of the themes in the book is to point out the unsuitability of managerial principles to direct science. The arguments are particularly emphasized in Chapter 5.

Chapter 2 describes the qualifications usually required to become a researcher. Chapter 4 proposes some of the attributes that research scientists need to develop to have a successful career.

I tried to include examples of my own experiences as a researcher to illustrate some of the points. I feel that this, in certain cases, is preferable to more abstract discussions. The careers of many people are influenced by chance events that occur during their lives. In my case, I happened to listen to a talk by Sir Peter Medowar, a Nobel Laureate, in which he outlined the views of Sir Karl Popper, one of the foremost philosophers of science. This made me realize the importance of acquiring an understanding of the scientific method for a career in research. Thus, Chapter 3 has been devoted to a discussion of aspects of the scientific method. Popper proposed several areas that pose a danger to scientific progress. One was a lack of motivation for inquiry. The period from the 1950s to the 1970s, was a pinnacle of excitement in science as described in Chapter 8. Many scientific and technological advances were made during this period. Since then, there are signs that some of this excitement has declined. This shows up in the falling enrollments in science courses and the decrease in science's share of high-achieving students, which has been observed in some countries. Hopefully, this will be a passing phase. It will be up to those who take up the profession to bring about a return of this excitement. There is no reason why this should not be achieved. Great discoveries are there, waiting to be made. All that is needed are the conditions for scientific research to be suitable. The creation of these conditions will depend on those who lead. Some thoughts about leadership in science are put forward in Chapter 6. We learn from history, and it is valuable to peruse the careers of individual scientists who have made an impact. A few examples of such scientists are discussed in Chapter 7, to help give insight into how they arrived at their achievements. The final chapter (Chapter 8) turns attention to two related challenges facing the advance of scientific research. The first is the need to free scientists from the constraints that prevent them from realizing their true creativity. The second, which is directly related to the first, concerns the need to achieve a more pleasant working environment for many scientists.

I feel that it is important that scientists do not take themselves too seriously, so at the end of some of the chapters, a few humorous anecdotes are included.

# The Author

Finlay MacRitchie was a professor in the Department of Grain Science and Industry, Kansas State University from 1997 to 2009. He is presently Professor Emeritus in that department. Prior to this, he was a research scientist in the Commonwealth Scientific and Industrial Research Organization (CSIRO) of Australia. He has spent short periods of time as Visiting Professor at the University of Chile and the Federal University of Rio de Janeiro, Brazil, and as Senior Research Fellow at the Agricultural University, Wageningen, The Netherlands; the University of Paris V; the University of Lund, Sweden; and the University of Tuscia, Viterbo, Italy.

MacRitchie has published more than 150 papers in refereed journals and two textbooks—*Chemistry at Interfaces* (Academic Press, 1990) and *Concepts in Cereal Chemistry* (Taylor & Francis, 2010). He is listed as an Institute for Scientific Information (ISI) highly cited researcher.

He has been a member of the editorial boards of *Advances in Colloid and Interface Science*, *Cereal Chemistry*, and *Journal of Cereal Science*. Currently, he is editor-in-chief of the *Journal of Cereal Science*.

MacRitchie's awards include the F.B. Guthrie Medal of the Cereal Division of the Royal Australian Chemical Institute (RACI) and the Thomas Burr Osborne Medal and George W. Scott Blair Memorial Award of the American Association of Cereal Chemists (now AACC International).

# CHAPTER 1

# Introduction

## Motivation and Requisites for a Research Career

Many who embark on a career in scientific research are motivated by a curiosity about the world and a desire to tackle and solve problems. This is one thing that comes through clearly when research scientists are interviewed. In a U.S. PBS (Public Broadcasting Station) television program, *Cool Careers in Science* (Scientific American Frontiers), discerning questions were posed to a number of notable scientists and engineers. The text can be accessed online and should make valuable reading for those contemplating a research career.* These researchers were asked questions that included why they decided to become research scientists, what academic background is needed, what they do during a typical workday, what they like most about the work, and if there was anything they do not like. There was a certain consistency in the answers, as may be expected, and some of the points made by the interviewees will be discussed throughout this book. There are many requisites for researchers, but I believe that there are three that are essential if a scientist is to have a successful career.

*Curiosity or a Spirit of Inquiry*
The first is to have a curious or inquiring mind, as already noted. This is what gives a person the incentive and persistence to pursue a problem, with the objective of gaining an understanding of it greater than what has been held previously. As stated by Malcolm Cohen, one of the Frontiers interviewees, "Sometimes I get to feel that I'm the only person in the world who fully understands a particular question."* This is a unique position to attain. When Albert Einstein was working on his relativity concepts, it is said that he became quite ill when he realized the enormous impact of what he had discovered.

---

* PBS (Public Broadcasting Station) (www.pbs.org/safarchive/5_cool/53_career.html), accessed July 2010.

### Deep Knowledge of at Least One Discipline

The second requirement is that the researcher gain a full and deep knowledge of the discipline that is involved or potentially can be applied to the problems that will be tackled. This usually means a deep understanding of a basic science such as physics, chemistry, one of the biological sciences, or engineering. Formal course work at universities merely provides a platform from which researchers must extend their knowledge by dedication to the scientific literature. Some graduate studies try to emphasize the need to develop breadth of knowledge. This means covering a smattering of knowledge in a large number of areas. The fear is that scientists will become too specialized if they concentrate on one narrow area. However, from the point of view of research, what is required is depth, not breadth, of knowledge. Research is about advancing new knowledge. It is illogical to think that someone can advance new knowledge without having acquired a level of knowledge similar to what has already been achieved in a particular discipline. Albert Einstein reached a point where he had to master some mathematical principles before he could progress with his work. This can happen in other areas of creative work, not only science. Beethoven struggled to master counterpoint before he could use it in his music. The concern about specialization, of course, is valid. This, however, may not be a problem. Most researchers, if they have a curious mind, will be interested in fields other than that in which they are specializing and will take steps to become informed about them.

### Intense Effort and Tapping into Infinite Intelligence

The third important requirement is intense and sustained effort. Frontier interviewees stress that research is not a nine-to-five job. It is not easy to create new knowledge. It requires a relentless pitting of the intellect against the unknown. When scientists apply intense effort to trying to understand a problem, a stage is sometimes reached where they are able to tap into the ether or infinite intelligence, a universal limitless form of knowledge that is accessible to people who learn how to capture it. This is a stage reached by creative individuals such as the great composers or those who actively use prayer. It may be familiar to those who have read the book *Think and Grow Rich*, by Napoleon Hill (1960). When this occurs in science, a special state may be reached where the mind becomes receptive to ideas and answers to questions that have been posed. Of course, a deep knowledge of the relevant discipline is needed, as mentioned above; otherwise, it will not be possible for worthwhile ideas to be received and assimilated, just as seed sown on arid land will not propagate. Scientists are usually not too interested in becoming wealthy, as described by Hill, but the principles involved in acquiring riches may be similar to those required for receiving creative ideas. One of the dangers in research which has been enunciated by some of the Frontier interviewees is that of maintaining a balance in life. According to Jim Cordes, "It is hard to find the right balance between doing a complete job and remaining competitive in the field while achieving satisfaction in other aspects of life."* This is a challenge and must be resolved by each researcher individually.

## Early Interest and a Simple Research Problem

My own interest in science arose in a similar way to many of the Frontier interviewees. I grew up on a farm on which dairying was the main activity. When pasture was first established on the farm, good yields of quality species were obtained. However, as time went on, deterioration occurred, and some grass species such as subterranean clover disappeared. The fields where this was observed were used for hay making—that is, the pasture was shut off from grazing during the growing season (spring/summer) and then harvested for hay to feed the stock during the period of low pasture growth (winter). An explanation for the loss of quality was the challenge. At the same time as I was thinking about the problem, I began to read articles in agricultural journals. My reading was primarily for interest, but nevertheless, I kept the problem in mind. Some ideas emerged from my reading. Removal of grass from a field without returning it (at least partially) by feeding it to animals in the same field could potentially produce a transfer of fertility. That is, nutrients could be drawn from the soil and, if not returned, could cause a deficiency of one or more of the elements needed. Plant growth requires a number of essential elements, such as nitrogen, phosphorus, potassium, and sulfur, and trace elements, such as copper, zinc, and molybdenum to name some of the main ones. Another observation made at the time was that

---

* PBS (Public Broadcasting Station) (www.pbs.org/safarchive/5_cool/53_career.html), accessed July 2010.

where tree logs had been burned in the field, grass later appeared to grow abundantly, including clover. I read in an article that this had been observed and that the ash from the burnt wood contained high amounts of potassium. A tentative hypothesis emerged that the loss of pasture quality was due to removal of at least one important element as a result of repeated hay cropping. A subsidiary hypothesis was that potassium could be the culprit. Every year, the fields were top dressed with superphosphate, a standard procedure at that time. It was surmised that deficiency of phosphorus and sulfur (which was contained in superphosphate) was likely not responsible for the poor pasture. In fact, it seemed that application of superphosphate may not have been contributing to improvement and may well have been a waste of money and time.

An experiment was designed in which a small part of the field was divided into measured plots and each plot treated with a different fertilizer, a standard procedure used by soil scientists to test elemental deficiency. The treatments included none (control) and all the essential elements that were known. At harvest time, one of the plots stood out from all the rest. The plot treated with potassium fertilizer produced a lush growth of pasture, whereas all the others appeared little different from the control. A more sound experiment would have involved cutting and weighing the grass from each plot. This was hardly necessary, so dramatic was the response to potassium. When the mower was put through this plot, it continually stuck and had to be stopped and cleared every few feet.

This example of a simple research project has been described as it illustrates some points that I believe are important in scientific research. It shows how curiosity is a vital ingredient. The capacity to be challenged by a problem and to be ready to pit one's intellect to trying to solve it is paramount. Subsequently, the problems that I have worked on have been much less amenable, but the principles encountered in this first relatively simplistic problem were similar.

## Importance of Combining Study with Experimentation

Another point that needs to be stressed is the importance of combining thinking and reading with experimentation. Some students begin their research career by carrying out a "literature search" prior to beginning their experimental work, very often under the misguided instruction of a supervisor. As a result, there is little or no critical evaluation of the literature. There are little criteria for distinguishing what is good science and relevant from what is not. Too often, we see students form mind-sets based on certain things that are read or are taught but may not be sound or applicable to the problem being pursued. On the other hand, when students have the problem or problems at the back of their minds and are engaged in experimental work, they are more likely to develop a critical attitude. They are more able to separate what is good science and relevant to their research from what is not sound and useful to them. Their reading, therefore, becomes more directed and selective. A literature search done in isolation is not directed and resembles setting off on a journey without having a clear idea of the desired destination.

# References

Hill, N. 1960. *Think and Grow Rich*, Fawcett Crest, New York.
Public Broadcasting Station, PBS.org, Scientific American Frontiers Archives, Fall 1990–Spring 2000, *Cool Careers in Science* (www.pbs.org/safarchive/5_cool/53_career.html), accessed July 2010.

# CHAPTER 2

# Scientific Training and Personal Development

Sixty years ago I knew everything; now I know nothing; education is a progressive discovery of our own ignorance.*

**Will Durant**

## University Qualifications

In earlier times, it was possible for amateurs to make useful contributions to science. Today, this is much more difficult, and university qualifications are usually mandatory for someone to become a practicing research scientist. Requisites for university degrees in science vary between countries and even between universities in the same country. The following remarks are therefore rather generic and largely based on my own experience.

## Bachelor's, Master's, and Doctoral Degrees

The sequence of qualifications is first a bachelor's degree followed by an honors or master's degree, and ultimately a doctorate. The first degree (the bachelor's degree in many countries, or its equivalent), is usually a three- or four-year course. It may be longer for an engineering degree, but science degrees will be the main focus here. In the first year of a pure science degree, students will choose several disciplines for study. In later years, this number is reduced, and at least one (and perhaps two) is selected for specialization. It may be chemistry, physics, geology, botany, and so forth. This is called a *major* and is studied at increasingly advanced levels throughout the degree. When they graduate, students should have

---

* I Wise Wisdom On-Demand (www.iwise.com/Gwbjc), accessed July 2010.

a sound knowledge of the fundamentals of the science chosen as a major. An honors degree may be taken in the final year of a four-year science course for outstanding students or in the year following graduation with a bachelor's degree. It will usually consist of course work at an advanced level plus a short research project that will be written up as a thesis or dissertation. The master's degree follows and normally requires two years of study after a bachelor's degree. The doctoral degree (Ph.D.) is usually completed in three to four years after finishing an honors or a master's degree.

## Research Ph.D. versus Combined Research and Course Work

In the transition from the bachelor's degree to the postgraduate master's and doctoral degrees, there is a fundamental divergence of the requirements in different countries. In some countries, such as Australia, New Zealand, the United Kingdom, and Germany, the Ph.D. is a pure research degree. This means that the student pursues a research problem for several years, culminating in a thesis. The thesis is then sent out for examination by at least three experts of international standing in the field of the research. The outcome is decided by a panel from the student's university based on the examiners' recommendations. The student may fail, be asked to carry out additional experiments and resubmit, revise certain sections of the thesis, or simply make editorial corrections. The last two requirements may be relegated to the student's supervisor to ensure they are fulfilled before the degree is awarded. This type of doctoral degree requires no course work and therefore no written or oral examination.

In some other countries, including the United States and Canada, the degree requires in addition to a research project and thesis, satisfactory completion of a considerable amount of course work with its accompanying examinations. Prior to being accepted as a candidate for Ph.D., the student needs to pass a Ph.D. preliminary examination. This may take different forms depending on the university department. It may simply consist of written examinations in the student's area of specialization. Alternatively, it may consist of a literature survey of a topic followed by preparation of a research proposal appropriate for submission to a funding body. The literature survey and the final proposal are each presented orally to the student's graduate committee. The committee usually consists of four faculty members, including the student's major advisor, at least one member being from an external department. After successful completion of the Ph.D. preliminary examination, the student is admitted as a candidate for Ph.D. and is eligible to present the Ph.D. thesis after a certain minimum time (sometimes eight months) has elapsed. After completion of the thesis, the student presents a defense of the work in public to an audience followed by an oral examination of the thesis by the graduate committee.

From the perspective of training to become a research scientist, I consider the pure research degree without course work to be superior. Beginning at the age of five, through primary, secondary, and tertiary education, students are asked to sit in classrooms and are fed information. It has been said that children are always going around asking questions but that society has an effective way of dealing with this problem. It is called education! By the end of the first (bachelor's) degree, a student should have acquired a sound basic knowledge of at least one scientific discipline. Further force-feeding of information together with tedious assignments may only serve to erode the spirit of inquiry that is essential for research. When graduate students begin to pursue a research problem, it is important that they be given a certain amount of freedom to concentrate on the problem. This entails experimental work with simultaneous self-study and focused thinking. The advantage of this is that the books and articles chosen for reading will be relevant to the research, and the study will be more efficient. Much of the compulsory course work, on the other hand, may not be useful and may tend to instill mind-sets. Furthermore, without course work, the student will more readily develop a critical attitude to the literature. What is read can be compared with what is found by experiment. It is important to distinguish between acquiring information and acquiring knowledge. Information is received and stored passively. Knowledge, in addition, requires an active input from the individual. When the information is critically evaluated and considered to be sound, then this becomes knowledge. When a lot of information is dispersed in course work, there is a greater tendency to accept what is taught, and minds may become less critical. There may be a greater tendency to accept and not to challenge. This can result in regimented thinking rather than the imaginative thinking needed in science.

## Imaginative and Regimented Thinking

Regimented thinking can often be detected, for example, when manuscripts are refereed or when research proposals are reviewed. A research proposal may be rejected because the literature review was incomplete or there was insufficient detail about an experimental technique, even though the principal investigator may have been the one who developed the technique. More imaginative thinkers would give greater credence to other criteria, such as the solidity of the basic science, the track records of members of the research team, and the innovativeness of an idea and its potential to lead to far-reaching consequences.

Another aspect to compare between a pure research degree and one involving course work is the quality of the research content. If the two types of degrees require about the same time, three to four years, then logically the student who is free of the burden of course work will have more time to devote to research. Good research needs time for thinking to mature, for critical experiments to be designed and carried out, and for sound conclusions to be reached. I have heard it said that graduate students should be made to come into the laboratory and work in between classes. Most research, however, requires continuity. Some experiments take long times, or the time for an experiment cannot always be predicted. It is important to distinguish between genuine research and the collecting of data to put into figures and tables. Much of what goes under the name of "research" these days is simply the latter, and that can certainly be done in between classes.

Some may say that many graduates with master's and doctoral degrees do not go on to have careers in research. This is true. Many, particularly those who join an industrial company, proceed to management positions. However, if they are in charge of scientific and technical staff, it is crucial that they have a sound understanding of science. The way to understand science is through gaining some experience in carrying out research.

## The Ph.D. Preliminary Examination

The other main difference between the two types of Ph.D. degrees (at least in English-speaking Western countries) is that in the system involving course work, an oral exam is included, both for the Ph.D. preliminary and the final defense. I have not been able to discover the origin of the Ph.D. preliminary examination. I would be surprised if it had been inaugurated by research scientists. Typically, the completion of one degree qualifies a student to proceed to the next higher degree. A bachelor's degree with honors or a master's degree might be expected to qualify a student for admission to a Ph.D. degree. This is provided that a student has demonstrated in his or her theses for these degrees, the capacity to carry out a good research project and to present the work in a dissertation. The Ph.D. preliminary examination, in many cases, seems to be another obstacle together with course work to prevent a student from carrying out a research project of some depth. Some may argue that a student who later takes a position in a university science department will be expected to apply for competitive funding to develop a research program. Thus, the preparation of a research proposal as required in some Ph.D. preliminaries is valuable. It might be questioned, however, if a student at an intermediate stage of a doctoral degree has the background required to devise a worthwhile research project. Once a student has acquired sound knowledge in his or her area, as should happen toward the end of the student's doctoral degree and in their postdoctoral years, it might be more useful for the student to attend training courses in the preparation of grant proposals. Such courses are available, sometimes provided by funding bodies. I have also heard it said that, in a Ph.D. preliminary, a student can be assessed on his or her ability to demonstrate creativity. Creativity is a term, however, that should be reserved for true innovative contributions to thinking. It is an attribute that I suspect might be expected to be developed from toward the completion of a doctoral degree onward. Thinking up wild ideas based on *ad hoc* premises is not what should be thought of as being creative.

## Research versus Collecting Data

Returning to the obstacles put in the way of students, there is a danger in shortening the time available for dedication to the main research that students may not fully appreciate what is required for genuine research. I have the impression that some Ph.D. graduates believe that research consists merely of collecting data to put into tables and figures. Research is much more. Its aim is to discover new and useful knowledge. If newness were the only aim, then we could imagine a project where measurements of the dimensions of all the rooms in the building could be made. Laser equipment could be used to make measurements to the nearest micrometer. Experimentally, it would be good science. Statistics could be used to estimate standard deviations and to calculate correlations between the different measurements. This would be information not previously known, but would it be useful? This example seems rather absurd, but many studies that go under the names of "research projects" are not much better in terms of discovering useful knowledge.

## Oral and Written Examinations

An examination based on oral questioning rather than only on the evaluation of a written thesis can be contentious. Admittedly, it allows the student to be quizzed on the spot and gives an easier way for examiners to clarify issues. However, it introduces a variable—the personalities of the students as well as the examiners. I have seen students choke up when confronted by several aggressive professors, while other students have been given an easy ride by sympathetic questioners even though the research presented may be mediocre. I have heard it said that the oral examination tests the capacity of students to think on their feet. But is it important for scientists to be able to think on their feet? Certainly it is an asset when presenting a research proposal to a committee or debating points at a scientific conference. However, thinking on your feet is not an essential requirement for a scientist as it undoubtedly is for a lawyer or a politician. Scientific research is more about reflective thinking that requires time and rarely depends on quick judgments on the spot.

## Limitations of Statistics

Sometimes, when I ask a student today how their results are coming along, I may get an answer such as, "I haven't looked at them yet. I still have a few more results to get. When I finish these, I will start to look over them." This approach is totally wrong. I recall that when I was a research student, some of my fellow students would come into the laboratory and plan their experiments for the day or for the next few days or more. The planned experiments were based on expectations of how the results would turn out, in other words, on a hypothesis. They were excited and could not wait to see whether the results confirmed their expectations. More often than not, they did not, so they had to go back to the drawing board and think afresh about how they would continue. The two different approaches I described essentially illustrate the difference between the inductive method and the hypothesis-deduction procedure that will be described in the next chapter. Those who simply set out to collect data will not have a clear focus, and the work will usually be pedestrian. When they have compiled all the data, they may not even look at the numbers but will apply a statistical package to determine whether the correlations they obtain between different variables are significant. If the statistical program shows them to not be significant at a certain level, they will very likely not pursue the relationships. They may accept the verdict of statistics. However, in some cases, if they had looked more closely at the data, they may have noticed some trends that, if they had investigated further, may have given insights into what was really happening. Too many students these days are slaves of statistics. It should be said that statisticians can give important advice to students to plan their experiments and often give generously of their time to do this. However, they are usually not research scientists, so students should not follow statistics blindly. It should also be remembered that most of the great discoveries in science have been made without the application of statistics.

## Accuracy: Are the Results What Are Intended to Be Measured?

Another problem with the more inductive method relates to the question of accuracy. It is essential that the measurements one makes are truly what one is trying to measure. Often, it can be that the measurements recorded are not what are intended to be measured. This can occur if there are defects in the experimental procedure.

Sometimes, the experimental error may be greater than the true differences one is attempting to show. Alternatively, it may be that the procedure used is simply not sufficiently sophisticated to be capable of making the measurement or that certain variables are not recognized or are ignored. This is where the experimenter needs to be perceptive and be able to recognize that there are flaws in the methods. As a result, these first measurements need to be scrapped, and flaws have to be corrected. It may mean that a number of initial measurements have to be rejected. Some will say that you should never discard measurements. This is not true. If the experimentalist realizes that they are not accurate, they must be dismissed and the procedure fine-tuned until confidence is gained that they are authentic. It requires being alert and observant while carrying out the experiment to detect any part of the procedure that may need to be altered. Thus, before the scientist has confidence that the measurements are what are trying to be measured, a progressive refinement of the technique may need to be performed. This attribute of being able to recognize when measurements are not accurate and to find ways of overcoming flaws in the technique is an essential one for research scientists to acquire. Those who do not have this attribute may accept the first results they obtain which may well be inaccurate. Many of the more complex instruments used in science today are "black boxes," and this sometimes makes it more difficult to assess whether the measurements obtained are really what scientists believe they are measuring. In earlier times, research students were often expected to construct their own equipment. This led to a greater understanding of how it worked and its limitations. Technicians may well develop greater skills in using instruments than the research scientists who direct the work. However, it is likely that the research scientists will be more conscious of the need for accuracy. They will go to greater pains to discover errors and to eliminate them, whereas technicians may proceed according to the prescribed method and may fail to detect when errors creep in. How does one know if the measurements being made are accurate? There is no simple answer. The process of detecting and eliminating errors in the experimental procedure is one way of approaching true values. Another way is to use at least one other experimental approach. If two independent procedures point to the same result, the scientist can have some confidence that the result is accurate but, of course, can never be certain. Another parameter besides accuracy (how close the measured values are to the true values) is the precision (repeatability). It is important to be able to repeat the measurements (replicates) and obtain closely similar values. However, some fall into the trap of associating good repeatability with reliability, without establishing that the measurements are accurate.

## Where To after Graduation?

For a career in scientific research, a degree equivalent to a master's or a Ph.D. is usually essential. The time spent as a research student and the years immediately following completion of the research degree is the formative stage of a successful research career. There is some excellent literature dealing with the pursuit of a research career. Rather than risking repetition of much of what has been written, I will refer the reader to some of what is available and restrict my remarks to some personal observations. *Building a Successful Career in Scientific Research: A Guide for Ph.D. Students and Postdocs* (Dee, 2006) is a treatise that comprehensively covers the field. Donald Braben (1994) has written a fascinating book, *To Be a Scientist: The Spirit of Adventure in Science and Technology,* based on his experience in the encouragement of innovative science. *One Hundred Reasons to Be a Scientist* (Abdus Salam International Centre for Theoretical Physics, 2004) includes contributions from many scientists and should make stimulating reading for those involved in research. *Moving On in Your Career: A Guide for Academic Researchers and Postgraduates* (Ali and Graham, 2000) provides valuable advice on how to enhance career prospects in academic research or lecturing. *Planning a Scientific Career in Industry* (Mohanty and Ghosh, 2010) is directed to those seeking to work in industry. Other books that are recommended reading are two by Donald Braben, *Pioneering Research: A Risk Worth Taking* (2004) and *Scientific Freedom: The Elixir of Civilization* (2008). These two books are discussed further in Chapter 8.

## Main Areas for Employment of Science Graduates

Before finishing their university studies, students need to be thinking about their future employment options. Advisors can often be helpful in giving direction, as they have usually developed relevant personal contacts. There are three main avenues for research scientists to follow. These are university science departments, government research agencies, and commercial companies engaged in research and development. University science departments have faculty members whose duties, in addition to teaching, are expected to include development of research programs. Research scientists at universities have the greatest freedom to choose and develop their own directions in their field of expertise. Next in terms of flexibility are government research organizations dedicated to solving problems of national interest and may include basic as well as applied research. Examples of national research organizations are the CNRS (National Centre for Scientific Research) of France and the CNR (National Research Council) of Italy. Germany has the Max Planck Institutes, Australia the CSIRO (Commonwealth Scientific and Industrial Research Organization) and New Zealand the Crown Research Institutes. The United States has many national institutes such as the National Institutes of Health (NIH). Many countries have research institutes under their academies of science, such as is the case in China.

Those employed by commercial companies are expected to focus on the company's objectives that are normally to create new or improved products or processes. If a graduate chooses to work in one of these three major areas, this does not mean that he or she needs to be restricted to that area indefinitely. There is scope for mobility between the areas. In fact, someone contemplating an academic career can obtain valuable experience by initially working in industry. The knowledge of how industry works, which is different than academia, and the contacts developed will be a great asset when it comes to initiating meaningful research and seeking funding to support it. Conversely, industry can gain by taking personnel from academia who may introduce a more fundamental approach to tackling problems.

## Internships as a Precursor to Employment

Internships are a way that students can gain valuable experience in a scientific workplace and help them to make choices about their future. They are offered by both academic and industrial organizations. Students need to research the opportunities that exist. Frequently, companies may send representatives to university departments to recruit potential internees. The representatives will give talks to describe the type of work done in their company and give feedback to inquiries. Students can request interviews in which they can put forward their credentials. Internships are offered at both undergraduate and graduate levels. For graduates, the company representatives evaluate students based on their academic performance up to that time, their communication skills, and their record of participation in activities at student, university, and even national and international levels. With that in mind, students need to prepare for these interviews so as to portray themselves in the best light to enhance their chances to be offered internships. If they are offered internships, which are usually for a duration of three to six months, they then have to convince their supervisors to grant permission to accept. Some supervisors are not supportive of students taking time off from research projects to work on something quite different. This is understandable as they have invested time and funding for the research. The break in continuity of the research can sometimes cause students to lose their momentum and waste time in refocusing on their project. Despite this, internships are usually positive for students' careers. In the case of an internship in industry, it allows students to experience the working environment in a scientific organization. If the internship is in another university department, possibly in another country, students are able to extend their expertise and acquire fresh ways of thinking, different from their home department. The opportunity to be exposed to and stimulated by different modes of thinking is an asset for a career in research. Many companies use internships to recruit promising graduates. Sometimes, the prospective employer, once he or she has selected someone, puts pressure on the student to join the company quickly and to commence work by a certain date. This can be unfair to the best interests of the student and the university department and can jeopardize the quality of the student's final research.

## Networking, Joining Associations, Conference Presentations

The scientific community is a closely knit fraternity. It is valuable for scientists to establish contact with colleagues who work in their area of expertise. There are opportunities to do this through meetings and conferences at local, national, and international levels. It is also valuable to correspond with these colleagues, and this can now be done easily through e-mail. Those who aim to advance their careers should join professional associations that represent their area of research and be willing to serve on their committees. They should also try to attend and make presentations at conferences. When traveling to attend conferences, opportunities should be taken to visit colleagues at their institutions in order to hold discussions and observe the facilities being used. The formation of strong bonds with fellow scientists around the world is one of the most rewarding aspects of a career in research. In some cases, it may be necessary to personally subsidize the travel. This should be considered as an investment in scientists' careers. If possible, they should give an oral presentation. This is considered to have more prestige than a poster presentation and gives the individual more visibility. It is a valuable way of advertising their research and becoming known among their peers. However, poster presentations are also valuable. Copies of the poster should be attached to poster boards so that anyone who is interested in the topic can take a copy away and read it at their leisure. As the copy has the contact details of the author (or authors), it facilitates interchange of information. If no copies are supplied, poster boards usually have envelopes for visitors to leave their business cards to indicate that they would like to have copies of the poster sent to them.

When giving an oral presentation, it is important to present a simple message. A common mistake of presenters is to show too many slides and try to cram in as much information as possible. This can be a turn-off for the audience. It needs to be realized that the work presented will likely be unfamiliar to many in the audience. Therefore, it is important to present it in a way that will make it interesting and easily understood. A good deal of thought needs to be put into the organization of the presentation so as to achieve this aim. It is not necessary to provide the level of experimental detail that is required for a published paper. The information on the slides should be concise, and in the case of text, it may be best to limit it to one or a few bullets that serve as talking points. The audience does not have time to read a lot of text, and if they do have to, this takes their attention away from what is being said. It is vital that the presentation be completed in the time allocated and that time is allowed for questions and comments from the audience. To achieve this, the talk needs to be rehearsed and adjusted, if necessary, to comply with the time limit. The answers to questions should be concise, and if unable to answer a question, it is better to admit this than to try to waffle. If asked to chair a session at a conference, it is important to become as familiar as possible with the work to be covered in the session. Should there be no questions at the end of a talk, it is the obligation of the chair to ask a pertinent question or to encourage some discussion. It is demoralizing for a presenter to be met with silence after giving a talk. When preparing a talk, presenters should try to put themselves in the position of a member of the audience for whom the topic may be new. The talk should be structured so that the message flows logically and the points made in each slide are easily assimilated. Helpful advice for giving presentations can be found in *Scientists Must Speak* (Walters, D.E. and Walters, G.C, 2010).

# References

Abdus Salam International Center for Theoretical Physics. 2004. *One Hundred Reasons to Be a Scientist*, Trieste, Italy.

Ali, L., and Graham, B. 2000. *Moving On in Your Career: A Guide for Academic Researchers and Postgraduates*, Routledge, New York.

Braben, D.W. 1994. *To Be a Scientist: The Spirit of Adventure in Science and Technology*, Oxford University Press, New York.

Braben, D.W. 2004. *Pioneering Research: A Risk Worth Taking*, John Wiley & Sons, Hoboken, New Jersey.

Braben, D.W. 2008. *Scientific Freedom: The Elixir of Civilization*, John Wiley & Sons, Hoboken, New Jersey.

Dee, P. 2006. *Building a Successful Career in Scientific Research: A Guide for Ph.D. Students and Postdocs*, Cambridge University Press, New York.

Mohanty, S., and Ghosh, R. 2010. *Planning a Scientific Career in Industry: Strategies for Graduates and Academics*, 2nd ed., John Wiley & Sons, Hoboken, New Jersey.

Walters, D.E., and Walters, G.C. 2010. *Scientists Must Speak*, 2nd ed., Routledge, London.

## ON HOW TO SHOW THE SPIRIT OF SCIENTIFIC ENTERPRISE (ORIGINALLY PUBLISHED IN *PUNCH*)

There are three principal groups of spurious scientific activity, which come under the heading Bandwagon, No Stone Unturned, and Fancy That. Take Chigwell, for instance, cozily installed in a well-equipped laboratory at Oxbridge University, but without an idea in his head. He reads in an obscure Italian journal that Fettuchini of Bologna has isolated a new anti-cancer principle from an extract of alfalfa. He immediately sets his whole organization to work on the same subject, repeating the original experiment over and over again with certain modifications.

Within a year he has a whole series of publications ready on the same subject and he begins to fire them like buckshot all over the world. The editors of the scientific journals find themselves in a quandary. Chigwell's work is tedious, repetitive and unoriginal, but the subject on which he is working is so important that it is hard to turn down anything remotely connected with it. So they take the line of least resistance.

As a result of this, Chigwell soon ends up with more publications than the man who made the original discovery. His name begins to dominate the literature. In any review of the subject he must be mentioned. If there is to be an international discussion he must be invited. He is after all, in many ways better value than Fettuchini.

He speaks English, he is an amusing orator, he has a knack for explaining technicalities in simple language—and he is always available. Very soon people forget the meager nature of his contribution. "Chigwell!" they say. "Yes, of course—the Cancer man. A brilliant fellow." From that moment onwards he has nothing to worry about. He can dig away happily at the same hole for the rest of his life.

No Stone Unturned is a much less ambitious game than Bandwagon. It is thought to have been taken over originally from scholars in the humanities, who have made most effective use of it to maintain themselves in subsidized idleness over the centuries.

No Stone Unturned rests on the assumption that no piece of knowledge no matter how apparently boring or useless, is too small to demand a lifetime's investigation. Indeed in certain academic circles, where the study of large issues is regarded as showy and journalistic, triviality in itself may hold a certain cachet. A biologist who dedicated his working life to the study of variations in the biliary passages of the codfish carries the same aura of scholarship as a historian who has immersed himself in the lesser-known texts of the Venerable Bede.

The other great research game, Fancy That, is an extension of No Stone Unturned into the more modern science of psychology and sociology. Fancy That consists essentially of a series of elaborate and time-consuming investigations designed to give irrefutable scientific proof to a cliché. The practice of this game on a wide scale has filled the world of scientific literature with papers designed to show once and for all, that animals learn more quickly if properly fed or that human beings are distracted by loud noises.

Another group of games is connected with publication. The object of all publication games is to gain the maximum credit from the minimum of useful work. An expert player can make a little research go a very long way indeed.

The original published paper is merely a beginning. After this it is possible to deliver the same results assembled in a slightly different way, at a variety of meetings and discussions, all of which are in due course printed and add in a most gratifying way to the author's bibliography. Letters can be written to learned journals, calling attention to minor additions to the work, and ultimately review articles covering the subject in question with particular reference to the writer's own work.

Specific games, the names of which are more or less self-explanatory are Priority, in which half-digested experimental work is published in letter form so as to establish a leading position in the field, and Chasing the Grant, in which similar publication is made with the object of pressuring some organization into financing further work.

Linked with publication games is the great traveling game Symposium. This has the merit of wasting not only time but money also. Symposium is a seasonal game. Every year in the spring and autumn, the great scientific migrations begin.

The scientists can be seen like a flock of birds settling on Vienna or Tokyo or San Francisco, sometimes in small groups, sometimes in enormous clouds of several thousand at a time. Their shrill cries can be heard reverberating through meeting halls and hotel lounges. What are they doing there? Nobody really knows. Certainly no new scientific information is ever given out at these international meetings. The papers read at the formal proceedings are never anything more than short summaries of previously published work. Indeed, it has been suggested that the only reason papers are read is to satisfy donors of travel grants and income tax authorities.

With any kind of luck, a conscientious scientist can retire from all useful work around the age of 40 and still convince himself that he is leading a busy and useful life.

Indeed, if he wishes for material success he'd probably be better advised to do so than to devote himself to the very chancy business of experimentation. He will soon find out that no man can serve two masters. Being an important scientific figure is an occupation in itself. It leaves little time for playing about in the laboratory.

Consider for instance, our old friend Chigwell at the climax of his career. Now Lord Chigwell OM. FRS, syrupy of voice and portly of physique, he steps from the door of his Belgravia house into his chauffeur-driven Rolls. He has a heavy day in front of him. A committee at the Ministry of Technology about the peaceful uses of atomic energy, a board meeting at International Chemicals, then after lunch he has to give a lecture at the Royal Society of Arts on the importance of population control. In the evening, dinner with Lady Pamela Berry and a quick appearance on the Frost Programme.

He casts his mind back with nostalgia to the good old days in the lab, when he (with, of course some help from Fettuchini) finally cracked the problem of the alfalfa principle. That was a really basic piece of work, not the kind of half-baked rubbish the young chaps are playing about with now. Ah, he thinks sadly to himself, they don't do work like that anymore ... I'm sorry to have to tell him ... they do!

CHAPTER **3**

# The Scientific Method

For a career in scientific research, it is important to acquire an understanding of what constitutes the scientific method. Unfortunately, many scientists devote little time to thinking about the philosophy of science, and even in science courses at many universities, the subject is not given high prominence. Of course, there is no rigid formula for what constitutes the scientific method, and how it is applied will vary in response to the nature of the research being undertaken. Nevertheless, there are certain principles that govern how a problem should be tackled scientifically.

## The Scientific Method, Empiricism, Induction

What is meant by the scientific method? In earlier times, one favored answer was that science is distinguished by its empirical nature (that is, it is based on observation and measurement). However, astrology is based on measurements of the positions of heavenly bodies but is not seriously regarded as a science, more as a pseudoscience. Another idea that enjoyed popularity in the early 20th century was that induction was the method of science. Put simplistically, induction entails gathering a series of observations and then analyzing them to arrive at a generalization.

## Karl Popper: Analysis of Early 20th Century Theories

Perhaps the greatest insight into the scientific method was developed by Sir Karl Popper whose main contributions are summarized in two books: *The Logic of Scientific Discovery* (1959) and *Conjectures and Refutations* (1992). Popper grew up in Austria and was influenced by the intellectual environment there during the period just following

World War I. He was challenged by the problem of what constitutes the scientific method or, more specifically, what is the criterion that distinguishes a scientific theory from nonscience or pseudoscience. At the time, several theories were being widely discussed. These included the theory of Adler on "individual psychology," Freud's psychoanalysis, Marx's theory of history, and Einstein's relativity. What struck Popper about the psychological theories of Adler and Freud was their apparent explanatory power. There seemed to be no instances of behavior that could not be explained by these theories. Every event could be rationalized as a "verification." This led Popper to suspect that this apparent capacity to explain everything, rather than being a strength, as their adherents claimed, could be their weakness. In the case of Einstein's theory, it was quite different.

One of the predictions of the theory of relativity was that light should be subjected to gravitation. To test this prediction, expeditions were sent out in 1919 to measure the apparent positions of distant stars when their light passed close to a heavy object (the sun) during a total eclipse and compare them with their normal positions in the night sky when their light did not pass near the sun. The measurements confirmed that the light was deflected, and the magnitude of the deflection was consistent with what was predicted from the theory. What was remarkable about Einstein's theory was its vulnerability. The results of the eclipse measurements would not have been expected on *a priori* grounds, and if they had been different, for example, even if the light had been deflected but not by the right magnitude, the theory would have been refuted. This led Popper to conclude that the criterion of a scientific theory is that it must be refutable (or falsifiable). Using this criterion, the psychological theories of Adler and Freud failed, because no experiment could be conceived that could refute them. This did not mean that Adler and Freud were not seeing many things correctly, simply that their theories did not qualify as scientific. In the case of the theory of Marx, the original theory was testable, but Popper asserted that it had been tested and refuted. However, in order to preserve it, its adherents had reinterpreted both the theory and the evidence in order to make them agree. In this way, they made the theory irrefutable and thus destroyed its scientific status. In the case of Einstein's theory, it was testable and refutable as shown by the 1919 experiment.

## Demarcation: A Criterion to Distinguish between Science and Nonscience

Popper's objective was not to have a criterion to decide whether a theory (or hypothesis) was true or false but to distinguish between science and nonscience (or pseudoscience). He referred to the problem of drawing a line between science and pseudoscience as the "problem of demarcation." The criterion of refutability was thus the solution to the problem of demarcation.

Popper summarized his conclusions as follows:

> It is easy to obtain confirmations, or verifications, for nearly every theory—if we look for confirmations.
>
> Confirmations should count only if they are the result of risky predictions; that is to say, if, unenlightened by the theory in question, we should have expected an event which was incompatible with the theory—an event that would have refuted the theory.
>
> Every "good" scientific theory is a prohibition; it forbids certain things to happen. The more a theory forbids, the better it is.
>
> A theory which is not refutable by any conceivable event is non-scientific. Irrefutability is not a virtue of a theory (as some people often think) but a vice.
>
> Every genuine test of a theory is an attempt to falsify it, or to refute it. Testability is falsifiability, but there are degrees of testability; some theories are more testable, more exposed to refutation, than others; they take, as it were, greater risks.
>
> Confirming evidence should not count except when it is the result of a genuine test of the theory; and this means that it can be presented as a serious but unsuccessful attempt to falsify the theory. (I now speak in such cases of "corroborating evidence").
>
> Some genuinely testable theories, when found to be false, are still held by their admirers—for example by introducing some auxiliary assumption, or by reinterpreting the theory *ad hoc* in such a way that it escapes refutation. Such a procedure is always possible, but it rescues the theory from refutation only at the price of destroying, or at least, lowering its scientific status. (Popper, 1992, pp. 36–37)

With this insight, Popper rejected induction as a valid scientific method. Scientific knowledge proceeded from theory to observation, not from observation to theory as the proponents of induction asserted. Instead of waiting for regularities to appear, scientists actively try to impose regularities on nature. This leads to critical tests (i.e., observations that are not arbitrary but are undertaken with the intention of testing a hypothesis by obtaining, if possible, a refutation). It is unlikely that anyone would have compared the apparent positions of stars during an eclipse of the sun with their normal positions if the theory of relativity had not been conceived. Of course, nature frequently resists the imposition of a conceptual view, and the hypothesis that follows from it is refuted. In science, it is important to realize that a theory can be shown to be false if experimental results fail to agree with what was predicted. Conversely, however, it can never be shown that a theory is true beyond any doubt. If the results of experiments fail to refute a theory, all that can be said is that the theory has been corroborated and may provisionally be held to be correct.

## Myths as Precursors of Scientific Hypotheses

I have used the words *theory* and *hypothesis* interchangeably until now, but it is usual to distinguish between them. A hypothesis is a possible interpretation of experimental observations and is normally used to guess (called a *conjecture* by Popper) at a general explanation that can then be tested by further experiment. Theory is a term reserved to describe an interpretation that has received a good deal of acceptance. It may have evolved from a hypothesis (or hypotheses) that has been tested and corroborated. Of course, a hypothesis cannot be created out of nothing. There must have been observations and previous thinking on which to base it. Beliefs that have not been scientifically tested may be thought of as myths. Myths are examples of pseudoscience and can be considered as more primitive forms or precursors of scientific theories. They are characterized by dogmatic thinking (that is, the tendency to try to verify our beliefs even to the point of neglecting observations that conflict with those beliefs). In contrast, scientific thinking introduces the critical attitude characterized by a readiness to change a belief in the light of evidence that is contrary, such as the refutation of a hypothesis. Thus, in the evolution of the scientific method, the critical attitude is not so much opposed to the dogmatic attitude but rather is superimposed on it.

## Exploratory Work Preceding Hypotheses

It should be realized that scientific research cannot be reduced to a simple process of forming hypotheses and testing them. An important component of research is the exploratory investigation needed before it is possible to arrive at a plausible hypothesis. A certain degree of understanding of the area of study has to be obtained. During exploratory work, scientists proceed to gather information. As new facts are uncovered, critical thinking may lead to conjectures that are evaluated for their validity. Frequently, these initial conjectures have to be dismissed if they fail to explain all the observations. At some stage, as a result of careful observation and focused thinking, the scientist may gain inspirational insight into the problem being tackled.

The hypothetico-deductive scientific method therefore begins with a conceptual formulation or hypothesis aimed at explaining or rationalizing previously unexplained or apparently unconnected phenomena. This is the creative step in scientific research. Next, the concept or hypothesis is subjected to criticism and to experiments designed to severely test it, because if the conjecture is true, certain consequences should logically follow. This is the deductive step. Experiments are carefully designed to attempt to refute the hypothesis. Should the experiments fail to produce a refutation, then the hypothesis is considered to be corroborated. That means it is accepted for the time being but with the proviso that it may still be refuted at a later date as a result of new observations or rational arguments. On the other hand, should the hypothesis fail to pass the tests, it is rejected, and a search for a new one is commenced. This is synonymous with the critical attitude embodied in the scientific method. However, where attempts to ignore the results or to prop up hypotheses are made, this corresponds to a dogmatic attitude, typical of nonscience or pseudoscience.

## Growth of Scientific Knowledge

Popper emphasized an important aspect of science—namely, its need to grow. The growth of scientific knowledge does not mean a continuous accumulation of observations but a critical examination of theories and their refinement or replacement by better ones. In this sense, science is unique in that, by its nature, errors are systematically criticized and, over time, corrected. In his deliberations on the progress of science, Popper posed the question of whether there is a danger that the growth of scientific knowledge will come to an end because the task will have been completed. The answer, he believed, was an emphatic "no" due to the infinity of our ignorance.

## Dangers to Growth: Lack of Motivation for Inquiry, Misplaced Faith in Precision, Authoritarianism

However, Popper did see dangers to the growth of scientific knowledge and specified three. These are a lack of imagination or real interest, a misplaced faith in formalization and precision, and authoritarianism in its various forms. There have been times throughout history when there has been a lack of motivation for inquiry. We have seen that a spirit of curiosity for understanding the natural world is essential for discovering knowledge. If that is not present in a society, knowledge can stagnate. Today there are certain pressures that tend to discourage genuine interest in discovering new knowledge. The pressure to publish, discussed in more detail in Chapter 4, tends to restrict some researchers to the extent where their priority is to scratch the surface of a problem in order to put together a paper rather than conduct an in-depth study. A preoccupation with measurements and precision rather than with concepts is also a consequence of more superficial investigations. Authoritarianism has been the enemy of the freedom of thought required in science throughout history. Those who questioned that the earth was the center of the universe were persecuted. The science of genetics in Russia was dominated for a long period by a dogma. Imposition of dogma is the opposite of the critical attitude of science. Where there are authoritarian influences, dogma tends to prevail, and scientific progress tends to stagnate. A form of authoritarianism that was not so obvious at the time of Popper's expositions is the managerial systems imposed on science in recent times. These systems with their well-defined hierarchical structures and command and control mechanisms are discussed in more detail in Chapter 5.

Oppressive regimes can but do not necessarily always exist in political systems, organized crime, and corporate organizations. They attract a certain type of person, one who seeks to assume power and control. These persons usually have little or no feelings for fellow humans; they are motivated by greed and power. If this climate exists in a scientific organization, the loser is science and the creative people who are responsible for its progress. For science to flourish, intellectual freedom is required. Those who wish to control, perhaps at times unwittingly, try to suppress this freedom. They regard it as dangerous. Creative people are a threat to those who would control. Throughout history, there have been the creators and the destroyers. The creators are not usually concerned with acquiring power. All they have is their creativity. If this is taken away, they are left with nothing.

## How Scientific Research Can Be Put Off Track Deliberately

We have seen that science is unique in that, by its nature, errors are systematically criticized and eventually corrected. There are two ways scientific research can be put off track. The first is by deliberately presenting false evidence that may mislead researchers—in other words, a hoax. There have been many examples of hoaxes, and we will look at one of the most famous: Piltdown Man.

### Piltdown Man

In 1912, fragments of a skull and jawbone were recovered from a gravel pit at Piltdown, a village in East Wessex, England. Many experts of the day believed the bones were those of a previously unknown form of early human. The finding of the remains has not been clearly documented, but Charles Dawson was the first to present the evidence, claiming to have been given the skull by a workman at the Piltdown gravel pit. At the time, there was great interest in the evolution of humans, and it was thought that there had been a gap in the fossil records between early hominoids and modern humans. The Piltdown Man was proposed by many to fit this "missing link." Subsequently, Dawson was accompanied to the site by Arthur Smith Woodward, who was in charge of the geological department at the British Museum, and they found further fragments. Woodward proposed that Piltdown Man was the missing link between apes and humans. The findings seemed to combine a human-like cranium (albeit with a brain size smaller than that of modern man) with an ape-like jaw. This tended to confirm the view that was most accepted at the time that human evolution commenced with the brain.

It must be said that, from the time of the discovery, skepticism was expressed by many. Then, in 1923, Franz Weidenreich, an anatomist, examined the remains and reported that they were a composite of a modern human cranium and an orangutan jaw. Although this was correct, the controversy continued, and it was not until 1953 that the newspaper *The Times* published evidence gathered by a number of experts proving that the Piltdown Man was a forgery. It had taken 40 years from its "discovery" to its definitive exposure as a fraud. The identity of the forger has never been completely determined, but Charles Dawson has been the main suspect. He had perpetrated various archaeological hoaxes prior to the discovery of the Piltdown Man. The Piltdown fraud had a large influence on research into human evolution. The belief that the human brain expanded in size before the adaption of the jaw to new types of food confused the issue. Acceptance of this despite contrary evidence caused a considerable waste of time and held back the theory of human evolution for several decades.

## How Scientific Research Can Be Put Off Track Unintentionally

The second way in which scientists can be misled is not deliberately dishonest. Scientific research, as described in Chapter 4, often consists in traveling along dead-end roads until it is realized that these are not the paths to truth. On occasions, these dead-end roads have resulted in a great deal of wasted time and effort for the scientific community. Examples of this have been described by Irving Langmuir (see Chapter 7) under the term *pathological science* (Langmuir and Hall, 1989). The appropriateness of this term has, however, been criticized by Bauer (2002). The characteristic symptoms of pathological science are that "People are tricked into false results by lack of understanding of what human beings can do to themselves by way of being led astray by subjective effects, wishful thinking or threshold interactions."*

---

* Langmuir, I. 1953. Quoted in: *Bad Science* (www.catchpenny.org/patho.html), accessed August 2010.

Some examples include Martian "canals," N-rays, polywater, water memory after infinite dilution of antibodies, and cold fusion. We will look at one example of this, the polywater controversy.

### Polywater

In 1966, Soviet scientists discovered and reported that when water was heated and condensed in quartz capillaries, it had properties very different to ordinary water. For example, it had a higher density, a viscosity about 15 times that of normal water, a boiling point higher than 100°C, and a freezing point lower than 0°C. The director of the Laboratory for Surface Physics at the Institute for Physical Chemistry in Moscow, Boris Derjaguin, presented these results in England at a "Discussions of the Faraday Society" meeting, referring to the material as "anomalous water." The topic gained interest, and in the following few years, several hundred papers were published on it. Some of these papers repeated the work and reported on properties of the material. Some believed that it was indeed a new form of water that had polymerized and dubbed the substance "polywater." A concern that arose was that if this form of water were a more stable form, then if it were to contact ordinary water, it would convert it into polywater, with disastrous consequences for the world's oceans and all life processes. Others were skeptical and questioned whether it was an artifact associated with contamination introduced by the method of preparation. One of the problems was that because of the way it was prepared, using fine capillaries, only minute quantities were available for experimental study.

An issue of the *Journal of Colloid and Interface Science* (Volume 36, Issue 4, 1971) contains 22 papers devoted to the topic of polywater and makes fascinating reading. Although some of the papers supported the authenticity of the phenomenon, many were critical and presented evidence that it was an artifact. As one example, Lauver et al. (1971) reported a study in which the liquid appearing in glass microcapillaries during experiments to make polywater was shown by mass spectrometry to contain silicone stopcock grease. When stringent precautions were taken to eliminate oil, grease, and other contaminants as well as the possibility of capillary contamination by film creep, no water with the alleged properties of polywater could be made. Subsequently, the scientists who had described the properties of polywater admitted that it did not exist. They had been misled by experiments that were poorly controlled and problems with experimental procedures. As these problems were resolved, evidence for its existence disappeared.

The topic of polywater is an extreme example of what can happen when scientists are put off track and induced to spend time and effort on work that turns out to be essentially unproductive. In these cases, the desire to believe in a phenomenon often overshadows the detachment needed to critically examine it.

On the other hand, we must be aware that a theory that has been discredited or becomes out of favor with the scientific community for some time may return to favor at a later date. An example of this is the theory of Continental Drift. This theory was proposed in 1912 by Alfred Wegener. At the time, it sounded fantastic and was not taken seriously until well into the 1960s. As evidence grew, the theory gained favor and was eventually accepted. Care is therefore needed not to assume that if a theory is "not accepted" at a point in time, that this is always a useful criterion for rejection.

## References

Bauer, H.H. 2002. "Pathological Science" Is Not Scientific Misconduct (nor Is It Pathological). *HYLE—International Journal for Philosophy of Science* 8(1):5–20.

Langmuir I. 1953. Quoted in: *Bad Science* (www.catchpenny.org/patho.html), accessed August 2010.

Langmuir, I., and Hall, R.N. 1989. Pathological Science. *Physics Today* 36–38. (Hall transcribed and edited the talk by Langmuir in the Colloquium at the Knolls Research Laboratory [General Electric] on December 18, 1953.)

Lauver, M.R., Wong, E.L., Stearns, C.A., and Kohl, F.J. 1971. Polywater Preparation and Silicone Grease. *Journal of Colloid and Interface Science* 36(4):552–553.

Popper, K.R. 1959. *The Logic of Scientific Discovery*, Routledge, London.

Popper, K.R. 1992. *Conjectures and Refutations*, 5th ed., Routledge, London.

# CHAPTER 4

# Attributes Required by Research Scientists

Simplicity paradoxically, is the outward sign and symbol of depth of thought. It is only when thought becomes clear that simplicity is possible. When we see a writer belaboring an idea, we may be sure the idea is belaboring the writer.

**Ancient Chinese Saying**

In Chapter 1, three important requirements for success in scientific research were proposed: a spirit of curiosity, a deep knowledge of at least one area of science, and an intense focus on the problem being addressed. In addition to these requisites, other attributes need to be developed, and we will examine some of these.

## Citations as a Criterion for Research Value

Popper pointed out (see Chapter 3) that science should be thought of not as an accumulation of information but a greater understanding resulting from new ideas or refinement of present theories. Unfortunately, today a lot of activity that goes under the name of research is simply the former. The pressure placed on scientists to meet goals tends to reduce the freedom and time for the deep thought required to develop theoretical concepts. In the evaluations of scientists, a high weight is placed on the number of published papers. In view of this, many scientists concentrate on accumulating data, as this leads to publication more easily and with less risk. However, publications that increase our understanding by creating new or improved theoretical treatments should carry the most prestige. The best scientific papers are those that influence the thinking of researchers in the field and affect the direction of future research. Some measure of the impact of a publication may be derived by determining the number of times it has been cited by other authors.

The availability of databases and new computer programs is facilitating these determinations. A study by the Institute for Scientific Information (ISI), reported by Hamilton (1990), on the top 10% of all scientific journals worldwide from 1981 to 1984 revealed that 55% of the papers received no citations in the first five years after they were published. An earlier ISI study of articles in the hard sciences published between 1969 and 1981 showed that only 42% received more than one citation. If it were assumed that a similar trend applied to papers in the 1981 to 1984 study, then it could be concluded that as many as 80% of papers published in that period were never cited more than once. Furthermore, self-citation, a practice in which the authors cite their own earlier work, accounts for between 5% and 20% of all citations. It seems unlikely that this trend in citations would have improved since that study was made. According to the ISI analyst, David Pendlebury, "The concentrated wisdom in the field is that 10% of the journals get 90% of the citations. These are the journals that get read, cited and have an impact" (cited by Hamilton, 1990, p. 1331).

## Conceptual Thought Required to Form Hypotheses

As discussed in Chapter 3, the first step of the hypothetico-deductive scientific method is to formulate a conceptual idea to unify previously unexplained observations. In order to form hypotheses, a capacity for conceptual thought is required. In this creative step, a scientist confronts a problem and conceives a solution for it in a way that has not been pictured previously. Although it involves creativity, a useful hypothesis has to be based on solid scientific foundations built up by previous workers, otherwise it will risk being *ad hoc*. This is where deep knowledge of a scientific discipline is essential. Occasionally, an inspired idea may come suddenly. However, this will usually happen only after a good deal of thought has been focused on the problem. The ability to think conceptually is a gift that comes to scientists who, having acquired sound knowledge in a field, apply themselves to think deeply to try to come up with a hypothesis that potentially will explain the observations that have been made. If it is a good hypothesis, it can also be used to make predictions that can be put to the test. The testing of the hypothesis then becomes the deductive part of the process, but even so, the design of experiments can also require creativity.

Experimental design is an important area of statistics. For example, if in an industrial plant we need to know the optimal conditions for making a product, where several variables are involved (e.g., temperature and concentrations of several reagents), a matrix can be drawn up in which each of the variables is systematically varied. Statisticians are able to devise an experimental design that will economize the number of experiments but will include a sufficient number to give a reliable answer. This is not the type of experimental design to which I am referring. When a hypothesis is formed, experiments to test it are often not obvious. Therefore, imagination is needed to design the experiments (e.g., a test of the theory of relativity described in Chapter 3 by measuring the positions of stars). The experiments need to be severe tests of the hypothesis and lead to possible refutations.

## Detachment

An essential quality for a research scientist to develop is the ability to separate or detach oneself from one's own theories. This is something that goes against human nature and cannot be learned from a textbook. It can be learned only through experience by forming hypotheses, having them refuted by critical experiments or rational arguments, accepting that they were wrong, and going on to formulate new hypotheses. I would like to recount my own experience in this regard. While I was a graduate research student, I obtained results that caused me to reach an interpretation that was different than what had been generally accepted. When I took this "new theory" to my supervisor, he listened patiently. He then suggested that if my ideas were correct, if I did a certain experiment, it should result in a confirmation. I thought at the time that it was unnecessary to do the experiment, as I felt that the evidence I had was sufficiently strong. However, I decided to do the experiment so that I could go to my supervisor and say "I told you so." Unfortunately, the results of the experiment did not turn out as I expected and, in fact, showed that my interpretation could not be correct. I was so devastated by this that I could not return to the laboratory for several days. Finally, when I did return and began thinking more about the problem, another way of looking at it emerged. I realized that this new way was much sounder than the initial one. This event was repeated several times in the early part of my career. On each successive occasion, it became easier to accept that my earlier hypothesis had been erroneous and to proceed to formulate a better one. I regard this as the most important lesson I have learned in research. Later in my career, I found it easier to view my theories with greater detachment. I also found that when I accepted that they were wrong, I was invariably able to move on and develop better ones.

It is easy to fall into the trap of feeling a sense of ownership of a hypothesis and being reluctant to give it up, even when it appears to be contradicted by experiment or rational argument. The effect is that there is no room for new ideas to enter, and creativity is severely impaired. In extreme cases, scientists feel that they are being personally attacked when their theories are criticized. Of course, criticism is an essential part of the scientific process. If it is valid, then the theory is rejected, and a search for a better theory is put into place. If, on the other hand, the criticism can be shown to not be justified, the theory becomes strengthened. No one is the owner of a theory, even though it is customary to ascribe many theories to different names. For example, the theory of relativity is usually referred to as Einstein's theory. Without wishing to detract from the merit of Einstein, who may have been the greatest scientist who has ever lived, the theory of relativity was there all the time. Einstein happened to be the first to come across it.

## Perseverance

Research can be a very hard taskmaster. It is not easy to wrest the secrets from Nature, even though once they have been wrested, it is obvious that Nature has not been deliberately secretive. It is simply that we as scientists have not known how to go about the discovery process. Before they do find out how to go about it, scientists may have to travel up many dead-end roads and experience considerable frustration. They may often reach a point where they feel that they should give up. This is exactly the point at which they need to resolve to go the extra mile. This does not necessarily mean that they should continue on the same obstinate path, but they may need to consider changing direction. Scientists need to have faith that effort is always rewarded. This is hard to accept if they have spent days or weeks or more on work that seems to have been a complete waste of time. However, it may be that this apparently wasted effort is needed to bring them to a new point at which they can move ahead. They may receive reassurance by reminding themselves that others experience the same disappointments and frustrations. However, in order to arrive at success, they must not give up but continue the struggle.

## Ethical Standards—Plagiarism

The aim of science is to search for the truth. Thus, scientists are expected to hold to impeccable standards of honesty. There are abundant opportunities to deviate from these standards, and there have been cases where the dishonesty of scientists has been discovered and documented. In addition to being a burden on their conscience, exposure can have drastic consequences for their careers and for the reputation of science in general. It serves as a stark reminder to others who might be tempted to be wayward. The rejection of data points was previously discussed in Chapter 2 and may or may not be dishonest depending on the circumstances. Obvious examples of dishonesty involve fabrication or massaging of data in order to fit the data to preconceived ideas. A deceitful practice, hopefully rare, is to present data taken from another person or other published papers without referencing the source. This comes into the realm of plagiarism. The most blatant cases of plagiarism are when material is copied verbatim without crediting the source. This is relatively easy to identify. In recent times, there has been an explosion in the number of scientific journals and the number of published papers. This is making it more and more difficult for reviewers and editors to identify when material has been copied. Furthermore, there are many forms of plagiarism that range from the copying of material verbatim to more subtle forms where it may not be absolutely clear if there has been a transgression. For example, a scientist might express an idea during discussion at a conference, and the idea may subsequently be published by another without credit. One practice that is clearly plagiarism and causes irritation for many scientists is that of taking over ownership of an experimental procedure. John Smith publishes a paper in 2005 using a technique published by another author in 1993, perhaps with a trivial modification. The 2005 paper cites the original reference (1993). Then, in 2010, he publishes another paper using essentially the same technique, which is cited as Smith (2005) without crediting the 1993 paper. Subsequent publications by the same and other authors credit the procedure to Smith (2005).

Another opportunity for stealing ideas occurs when manuscripts are submitted to journals or research proposals are sent to funding bodies for review. William Lipscombe, a Nobel Prize winner in chemistry in 1976, stated that "I no longer put my most original ideas in my research proposals, which are read by many referees and officials. I hold back anything that another investigator might hop on and carry out. When I was starting out, people respected each other's research more than they do today, and there was less stealing of ideas."*

Early in my career, I submitted a paper to a journal, and it was rejected. A few months later, a paper was published which was close to a repeat of my paper. I suspected that the author may have been a reviewer of my paper, but, of course, I could not be sure because most journals guard the confidentiality of reviewers. In this case, there was no question of the idea being stolen, as there had been insufficient time for the work to be done and written up. However, rejection of a rival paper does allow priority to be established. In my case, I eventually did manage to get the paper accepted after a vigorous defense of the criticisms that had been raised, particularly by one reviewer. Although it was published later than the rival paper, it had an earlier date of submission. Unfortunately, this sort of rivalry between scientists is fairly common. The ideal situation is where there is an honest exchange of information between those working on similar problems. Frequently, scientists decline to review a paper or research proposal if they believe there may be a conflict of interest, and this is an honorable thing to do.

For a more complete discussion of ethics in science with a large number of case studies, *Scientific Misconduct: Definition and Real-Life Case Studies* by J.G. D'Angelo (in press) is recommended.

## Publication

An essential requirement for scientists is that they communicate to a wide audience any significant new knowledge that they discover. This is done by publishing papers in scientific journals, of which

---

* Bauer, H.H., Ethics in Science, Virginia Tech (www.files.chem.vt.edu/chem-ed/ethics/hbauer/hbauer-intro.html), accessed August 2010.

there are many, each specializing in a restricted area of the scientific spectrum. There seems to be no purpose in a scientist making an advance in knowledge if no one is made aware of it. Having recognized this, it is also true that there can be abuse of publication. Many papers are published today not with the aim of disseminating knowledge but to enhance résumés and increase promotion prospects. Some confirmation of this is gleaned from analysis of the number of citations for published papers, referred to earlier in the chapter. When I was a graduate student, there was a faculty member who divided scientists into two groups: those who did research and those who published papers. This was somewhat cynical and may have been a bit tongue in cheek. Nevertheless, there was probably some truth in it. Number of publications is one of the criteria most used for evaluation of scientific performance. Frequently, those who make the judgments are not familiar with the research, and the only thing they can go on is the number of publications. Once I had a conversation with a senior manager in my research organization about someone who had published a remarkable number of papers in the space of a few years and who was being considered for accelerated promotion. He told me that this scientist's productivity was astounding. When I asked him if he had read any of the papers, he replied that he had not but reiterated that this scientist's productivity was astounding. I did not pursue the matter further. I had read some of the papers and was in agreement that the number of publications was astounding, but I had serious reservations about using the word "productivity" so loosely.

It is usually considered that for scientists to be making satisfactory progress, they need to publish a certain number of papers in a year (say two or three). There was one occasion when I published only one paper in a calendar year. As a result, my evaluation was quite a bit lower than normal. It so happened that one journal had been unusually slow in publishing papers after they had been accepted. Thus, I had two papers published in January of the following year and one in March. I do not believe that my productivity had declined. The one paper was an anomaly. The person doing the evaluation had no way of knowing this. He had many evaluations to complete, and the only way for him to make an objective decision was to rely on the number of papers published in the pertinent calendar year. I mention this to warn younger scientists to be streetwise and take steps to try to insure that they publish a respectable number of papers in each calendar year, if that is the criterion used. They may not like the rules, but these are the ones that they need to play by. The fixation on the number of publications tends to consider all publications as being equal. The quality tends to be ignored, even though there may be chasms between the best and the worst.

For someone of principle wishing to build a reputation as a research scientist, the aim should always be to produce publications that present genuine advances in knowledge. Good scientists concentrate on carrying out a worthwhile research project. At a certain time, it becomes evident that the work has reached some sort of completion. Then, the time will be ripe to write up the work and submit it for consideration by a journal. Writing of the paper will fall into place easily and logically. It may be that during the writing, additional experiments are suggested to help make the message more complete. At the opposite end of the spectrum, scientists will plan to produce a certain number of publications in a given time. Often, the papers are practically written before the work is done. This research approach resembles the inductive method mentioned in Chapter 3 and considered not to be a valid scientific procedure. Unfortunately, a proportion of the scientific community is not sufficiently perceptive to recognize this and, based on the regimented thinking discussed in Chapter 2, tends to treat all papers as similar in quality. In this way, mediocre research is allowed to flourish without the criticism that is merited. Many papers are published these days with a large number of authors' names. Contributions by some of the authors are minimal. This is especially so where some in high positions in the hierarchy insist on having their names on papers emanating from their institute, even though they may not have contributed anything. For some papers, a large number of authors is justified; for example, in the elucidation of a genome, the task is enormous and can only be practically carried out by contributions from many. However, there are other multiauthor papers where it is hard to imagine what contributions have been made by each author. Some supervisors insist that their research students publish a certain amount before they can graduate. In some countries (an example is Sweden), doctoral candidates are expected to submit several published (and accepted) papers together with their dissertation. This system appears to work well as students are not so constrained by time limits and are able to produce research of a mature level. In systems where students have to finish in a limited time, the requirement to publish papers tends to belittle the value of research. It often leads to pedestrian papers that add little, if anything, to genuine knowledge. They are destined to join the large number of published papers that are never cited or even sighted, as mentioned earlier in the chapter.

*Scientists Must Write: A Guide for Better Writing for Scientists, Engineers, and Students* gives advice for scientific writing and emphasizes ways in which writing is important to scientists (Barrass, 2002).

## Service: Peer Reviewing

As scientists move ahead in their careers, they will increasingly be expected to carry out duties apart from their research. For example, they will be expected to serve on committees and may be assigned responsibility for certain tasks such as supervising equipment that is shared between different individuals or groups. The way that these duties are carried out will therefore impact on colleagues. They may also be expected to act as chairs for seminars or conferences and to organize symposia. I will make some comments on two areas of service in which scientists can expect to be involved—peer reviewing and serving on award committees. Each of these duties encompasses ethical principles.

As described above, scientists are expected to communicate their research to their peers and do so by submitting articles to journals. Of course, it is not just a matter of sending off an article and expecting it to be published. Scientific standards must be upheld, and so the editor of the journal invites reviewers (normally two) to evaluate the quality of the submission. The reviewers, who are chosen for their expertise in the area, are asked to give a recommendation. It is rare for a paper to be accepted unchanged. First, it may be rejected by the editor without review because it is inappropriate for the journal, may have obvious deficiencies in the science, or may fail to present any worthwhile new knowledge. Before submitting a paper, the author (or authors) should read the guidelines that each journal provides to ensure that the paper complies. If the paper is sent out for review, the editor will make a decision based on his or her own and the reviewers' evaluations.

Reviewing a paper is a serious task. Frequently, the topic will not be exactly in the area of the reviewer's expertise and may require many hours of work to become familiar with the subject and to be in a position to make a sound assessment. All scientists depend on reviewers, who serve voluntarily and are not recompensed directly. However, at universities and scientific organizations, reviewing is accepted as an important duty, and scientists are expected to allot a certain percentage of their time to it. Younger scientists who have had some experience in research and have published papers are well advised to volunteer their services as reviewers to journals. If they build up a record of accepting invitations to review and do this reliably, they may be invited to serve on the editorial board of a journal. This will give them prestige and help them along their career path.

In the main, scientists submit to accepting a moderate number of papers to review. Of course, journal editors should not reward good reviewers by overburdening them with invitations to review. At the other extreme, there are the freeloaders. These are people who expect their submissions to be promptly reviewed but continually refuse requests for them to act as reviewers. When asked to give a reason for declining, some say they are too busy. This is a cop-out. Everybody is too busy, but many scientists accept a heavy workload of reviewing, not because they are not busy and have a lot of free time, but because they take their responsibilities seriously. Of course, there will be occasions when a genuine reviewer is unable to comply with the invitation. In these cases, the reviewer will reply promptly and may try to suggest one or more alternative reviewers. With the great advances in computer programs to monitor, it is relatively easy for journals to obtain lists of reviewer records. I recently noticed that one scientist, who is well known in the field, had accepted to review one paper and declined 13. This is a person who is evidently highly regarded by peers and has received at least one award that I know of. Perhaps it would be fitting if this sort of information is considered before giving awards. There are other abuses of ethical principles related to reviewing. For example, some never reply to invitations or decline after several reminders have been sent. Others accept the invitation to review but never deliver. All this means that some authors are unfairly subjected to longer waiting periods than they should experience. Another reviewer may return a recommendation to accept without providing any comments or very few. Unfortunately for this reviewer, the other reviewer, who did the job conscientiously, may come up with a long list of obvious objections to the paper. This demonstrates that the first reviewer has taken the easy way out. Fortunately, I do not believe that this happens often, but it does happen.

## Service: Serving on Awards Committees

Scientists may be nominated for awards to recognize their outstanding research or other contributions, such as service to associations. To administer awards, it is usual to appoint committees. Members of award committees are entrusted with the responsibility to select the most deserving candidate from those who are eligible. If there is a perception that this is not achieved, it can lead to cynicism and to lowering of morale among the scientific community. The committee normally consists of a chair and several members, who ideally should be selected for their deep knowledge of the field and their capacity for sound judgment. The guidelines for prestigious awards are usually clearly defined, and it is the duty of the committee to make a choice of recipient based on these guidelines. Customarily, it is only possible for individuals to receive awards if they are nominated. Therefore, it is imperative that the committee be proactive to ensure that all the names of those deserving of the award are put forward. Members of committees are normally not eligible to nominate. However, they have the responsibility of encouraging others to nominate deserving candidates. It is highly unfair that anyone who is eminently deserving of an award fails to be considered because he or she is not nominated.

I have observed that some award committees are not proactive in this sense. They may give little thought to potential candidates, wait passively for nominations to be made, and then perhaps select the candidate they think is best. This often leads to awards being made every year. There are two problems with this approach. First, it can lead to deserving candidates not being nominated. Second, recipients should only be selected if they fulfill the guidelines of the award. It is not simply a matter of choosing the best candidate. If none of the nominees meet the criteria outlined in the guidelines, then no award should be made.

What can happen if a committee is lax is that the selected awardee may simply be the one most strongly supported by friends and colleagues. It can also be that there is only one candidate nominated. The best person to nominate a candidate may well be a colleague from the same institution. Unfortunately, because of apathy or petty jealousy, a nomination may not be made. This is where the influence of the award committee is essential.

The choice of members of award committees is crucial. It should not be made to reward friends or to include the right balance of minority groups. It is such a serious responsibility that only those known for their soundness of judgment should be considered. To submit a nomination, the nominator provides a letter outlining how the nominee meets the criteria of the award and reasons why he or she should be a recipient. In addition to the nominating letter, it is advisable to include or request supporting letters from other distinguished scientists who are familiar with the work of the nominee. The committee also has the choice to seek independent assessment of the candidate from others with expertise in the field. This can be valuable for obtaining more objectivity, because supporting letters are always highly positive, as they should be.

## Grantsmanship

The tightening of budgets at universities and research organizations means that scientists are increasingly being asked to find funding for their own research programs. There is a limit to these funds, so there is intense competition and many miss out on a share. The pursuit of funding has become a serious activity for scientists trying to develop research programs. Searching for funding sources and writing and submitting proposals can be very time consuming. It is quite possible for scientists to turn it into close to a full-time activity. This time then encroaches into the time needed to carry out other duties, such as teaching, peer reviewing, serving on committees, carrying out research, supervising graduate students, and writing papers. The time devoted to seeking funding therefore needs to be planned efficiently. For example, university faculty members may have more time to devote to grant writing during the long vacation when the teaching load is reduced. It is important to understand what is required in preparing a proposal. Some funding bodies organize workshops in which guidelines are given for writing successful proposals. Attendance at these is highly recommended, although it may be necessary to travel. In preparing proposals, attention needs to be focused on closely adhering to the instructions and submitting a proposal that falls within the priority areas of research that are invited. Some funding bodies have panels responsible for specific areas of research. It is advisable before submission of a proposal that applicants contact the panel manager to ensure that the topic they intend to propose and its costing are suitable. I know of one instance where the funding body issued new guidelines in which they stipulated that larger projects having budgets with a higher range than previously stated should be submitted. I happened to meet one of the panel members at a conference after decisions had been made. He said that when the panel met, their priority was to fund as many projects as possible. They therefore began their deliberations by eliminating those proposals with the highest budgets. These happened to be in the range recommended by the funding body administration. There was thus an obvious mismatch between the issued guidelines and the *modus operandi* of the panel. Funding bodies normally try hard to remove any conflict of interest in evaluating proposals. On one occasion, I discovered by accident that one of the reviewers for my proposal was someone who had also submitted a proposal to the same panel. This can happen because there are only a limited number of experts who are qualified to act as reviewers for a given proposal. To deal with these sorts of issues, it is important to maintain a good line of contact with the panel manager.

Successful grant applications usually have a low success rate. Some large funding bodies have sufficient resources to fund only about 10% of proposals. Thus, applicants have to become accustomed to many failures. Large funding bodies usually provide reviewers' reports to applicants. If the first application is not successful, applicants need to revise their proposals and make improvements to address all of the reviewers' comments. These improvements should be clearly stated in the documentation submitted with the new proposal. If this is done conscientiously, the chances of success will be enhanced the second or even third time around. It is always helpful to include preliminary results in any proposal, whether it is the first or a revised one.

Funding may also be sought directly from industrial companies. Faculty members may act as consultants or obtain funding for a project to benefit the company. This may take the form of financially supporting a graduate student for a time. To initiate a project, the scientist may visit the company and present a seminar outlining the expertise that can be offered to the company. If the company is interested, a confidentiality agreement will need to be signed so that the company can divulge information. If an agreement is reached for funding, it is imperative that the scientist become familiar with the problem to be tackled and not accept without question what he or she is told it is. This will often take the form of visits to the plant to become acquainted firsthand with what is involved. Sometimes company representatives will have their own interpretation of the problem. This may not correspond exactly with the perception of the scientist after visits to the production line and discussions with company staff. Also, the company may have staff who suffer from the regimented thinking that was mentioned in Chapter 2. They may be biased against some sound approaches and have fixations about other dubious ones. The persons seeking funding support may need to be assertive in defending their views and explaining clearly the flaws in the alternative approaches.

A new appointee to a faculty position in a university science department may consider a mixture of fundamental and applied research for his or her program. Research has become more multidisciplinary. To be successful, it is almost essential that new faculty members look to colleagues in other departments to develop cooperative research projects. Multidepartmental and multidisciplinary teams tend to be favored by funding bodies. There are often special grants for new faculty or younger scientists, and these should be actively sought. When developing a research program, a younger scientist should aim to build it on solid foundations. This means that the structure should be based on science that has been tried and tested. It is preferable to build slowly and surely rather than rush into a program based on shaky premises. Like poorly constructed buildings, research programs, if not firmly based, can become unstable and collapse.

# BRAINSTORMING

The requirement for deep reflective thinking in research has been emphasized. Some science departments organize frequent meetings for research staff. These can be valuable, particularly in industry where, often, multidisciplinary teams are engaged in projects, and it is important to keep members of the project updated. These meetings, if properly organized, will be short. In more academic surroundings, meetings are often unnecessarily frequent and excessively long. The reason that they are over long is that there are always those who thrive on meetings as it gives them the opportunity to exercise their natural verbosity. It is also an opportunity for some to engage in bullying tactics, point scoring, and one-upmanship. Those who are not naturally verbose find this activity difficult to cope with. It is not what they expected when they embarked on a research career. I have known postdocs and younger scientists that could not support such an environment and have left to find other employment. Often, these are talented scientists who the organization can ill afford to lose. One aim of meetings is to enable those who have no ideas of their own to extract ideas from those who have. Those who are alert to having their ideas stolen may be reluctant to provide them and thus run the risk of being branded as not being team players. Another aim is to get information without the trouble of having to read the literature. The latter is a dangerous activity, as it may lead to absorbing facts that are not reliable. One of the ploys of meetings is to use them for brainstorming sessions. Brainstorming is an activity that appears to have begun in advertising companies where the aim was to develop jingles that sell products like soap powders.

The way brainstorming works is that people form a ring around a table. To begin, one person says the first thing that comes into his or her head. They then go around the table, each having an input. The idea is that, in this way, they will eventually arrive at the absolute truth. It is a bit like the theory that if you put a group of monkeys in a room with a piano and reams of sheet music, providing they are left there long enough, they will eventually compose all the symphonies of Beethoven.

## WHY I BECAME A SCIENTIST

My decision to pursue a career in science was mainly influenced by terrific professors at Vassar. I entered college as an English major, intending to become a writer. A freshman-year course in personality psychology changed my direction.

At Vassar, I worked as a research assistant for a professor studying memory. The training I received in my biology, chemistry, and physics classes sharpened my skills. I even volunteered as an assistant in my organic chemistry professor's lab, studying the properties of amber. This research had nothing to do with my career plans, but I loved the excitement of scientific discovery.

After 30 years as a college professor, I have come to believe that the key to inspiring students to pursue careers in science is not making science appear relevant to everyday life but helping students experience the excitement of the research enterprise. Few things are as satisfying as constructing a hypothesis, designing an experiment to test that hypothesis, carrying out the experiment, and discovering whether you were right.

Too much of today's science education focuses on making students memorize bits of information that will be outdated within a few years. Too little emphasizes how to think like a scientist. And there is no substitute for hands-on research experience.

Laurence Steinberg, Distinguished Professor of Psychology, Temple University, Philadelphia (Steinberg, L., December 2006/January 2007, "Science in the Spotlight," *Science Educational Leadership*, 64(4) (www.ascd.org/publications/educational-leadership/dec06/vol64/num04/Why-I-Beca). Copyright 2006 by the Association for Supervision and Curriculum Development.)

*My mother made me a scientist without ever intending to. Every Jewish mother in Brooklyn would ask her child after school "So? Did you learn anything today?" But not my mother. "Izzy" she would say, "Did you ask a good question today?" That difference—asking good questions—made me become a scientist.* (Isidor Isaac Rabi)

*Science is facts; just as houses are made of stones, so is science made of facts, but a pile of stones is not a house and a collection of facts is not necessarily science.* (Henri Poincare)

*At Princeton, Albert Einstein was more like a kindly uncle. When he arrived in 1935, he was asked what he would require for his study. He replied, "A desk, some pads, a pencil, and a large wastebasket to hold all of my mistakes."*

*The real scientist ... is ready to bear privation and, if need be, starvation, rather than let anyone dictate to him which direction his work must take.* (Albert Szent-Gyorgi)

*It is a good morning exercise for a research scientist to discard a pet hypothesis every day before breakfast. It keeps him young.* (Konrad Lorenz)

*In science it often happens that scientists say, "You know that's a really good argument; my position is mistaken." And then they actually change their minds and you never hear that old view from them again. They really do it. It doesn't happen as often as it should, because scientists are human and change is sometimes painful. But it does happen every day. I cannot recall the last time something like that happened in politics or religion.* (Carl Sagan)

---

* BookRags.com (www.bookrags.com/quotes/Isidor_Isaac_Rabi), accessed April 26, 2011.

† The Quotations Page (www.quotationspage.com/quote/33017.html), accessed April 26, 2011.

‡ Brain Athlete (www.brainathlete.com/category/thinkeinstein/page/3/), accessed April 26, 2011.

§ BrainyQuote.com (www.brainyquote.com/quotes/.../a/albert_szentgyorgi.html), accessed April 26, 2011.

¶ The Quotations Page (www.quotationspage.com/subjects/science), accessed April 26, 2011.

** The Quotations Page (www.quotationspage/com/quote14337.html), accessed April 26, 2011.

*I am among those who think that science has great beauty. A scientist in his laboratory is not only a technician: he is a child placed before natural phenomena which impress him like a fairy tale.* (Marie Curie)

*It is of interest to note that while some dolphins are reported to have learned English—up to 50 words in correct context—no human being has been reported to have learned dolphinese.†* (Carl Sagan)

*The doubter is a true man of science, he doubts himself and his inter-pretations, but he believes in science.‡* (Claude Bernard)

* The Quotations Page (www.quotationspage.com/quote/34022.html), accessed April 26, 2011.
† Vacilando (www.vacilando.net/node/273221), accessed April 26, 2011.
‡ FinestQuotes.com (www.finestquotes.com/author_quotes-author-Claude%20Bernard-page-0.htm), accessed April 26, 2011.

## PARTICULAR UNCERTAINTY

Despite Werner Heisenberg's Nobel Prize for its formulation, Albert Einstein never accepted the so-called "uncertainty principle" (which states that the more carefully one measures the position of a given particle, the less certain its momentum becomes) because it threatened to wreak havoc with the strict determinism in which he believed.

Indeed, the uncertainty principle was a subject about which Einstein and Niels Bohr argued many times over the years. On one memorable occasion (at the Solvay Conference in Brussels in 1930) Einstein unveiled the product of one of his famous "thought experiments"; an imaginary device comprised of clocks and scales, which, he claimed, violated the principle.

Following a sleepless night, however, Bohr discovered that Einstein had made a critical error; he had neglected to take into account the fact that clocks run slower in a gravitational field, a consequence, rather ironically, of Einstein's own theory of relativity.

Little Brown Book of Anecdotes (www.anecdotage.com/index.php?aid=11843), accessed April 26, 2011.

*If we knew what we were doing, it would not be called research, would it?** (Albert Einstein)

*May every young scientist remember and not fail to keep his eyes open for the possibility that an irritating failure of his apparatus to give consistent results may once or twice in a lifetime conceal an important discovery.*† (Patrick M.S. Blackett)

---

* QuoteDB.com (www.quotedb.com.quotes/2310), April 26, 2011.
† QuotesBy.com (www.quotesby.co.uk/quotes/q113291), April 26, 2011.

# References

Barrass, R. 2002. *Scientists Must Write: A Guide for Better Writing for Scientists, Engineers, and Students*, Routledge, New York.

D'Angelo, J.G. (in press). *Scientific Misconduct: Definition and Real-Life Case Studies*, Taylor & Francis, Boca Raton, Florida.

Hamilton, D.P. 1990. "Publishing by—and for?—the Numbers." *Science* 250:1331–1332.

# CHAPTER 5

# The Impact of Managerialism

My grandfather once told me that there were two kinds of people: those who do the work and those who take the credit. He told me to try to be in the first group, there was much less competition.*

**Indira Gandhi**

## The Managerial Ideology

In the past few decades, a rise in the influence of managerialism on scientific research has been seen. Those entering the field must expect to come up against this reality. Briefly, this change concerns the application of business management principles to organizations with the aim of producing economic efficiency or, put another way, the pursuit of maximum output with minimum inputs. A central dogma of managerialism is the belief that organizations have more similarities than differences; thus, the performance of all organizations can be optimized by the application of generic management skills and theory. To those practicing this system, there is little difference in the skills required to run a research organization, an advertising agency, or a factory turning out machines. The experience and skills associated with the organization's core business are considered to be unimportant or at least secondary. This management ideology has been applied to many organizations, but I would like to focus on the effects that it has had on scientific research and will take as an example the Commonwealth Scientific and Industrial Research Organization (CSIRO) of Australia.

---

* The Quotations Page (www.quotationspage.com/quote/39781.html), accessed August 2010.

## Commonwealth Scientific and Industrial Research Organization (CSIRO) of Australia

The CSIRO was established under an Act of Parliament in 1949, being an extension of the previous CSIR (Council for Scientific and Industrial Research). Its aim was to provide research for industry and government to enhance economic development. It was structured into divisions and smaller groups or units, each focused on a specific industry, and was headed by an executive management council consisting of a full-time chairman and four full-time members with a small number of part-time members drawn from business and university backgrounds. All the full-time members were senior scientists and employees of CSIRO. Divisions were headed by chiefs, and units by officers-in-charge who reported directly to a member of the executive management council. In a few decades following its formation, the organization flourished and made important contributions to Australian industries as well as to fundamental research such as radio astronomy. Later, its activities expanded into other areas affecting the community, including environment, human nutrition, urban and rural planning, and water supplies.

## Reviews of CSIRO

CSIRO has been subjected to numerous reviews since its establishment, but I will mention only two that have resulted in appreciable reorganization. As a result of the 1977 report by the Birch Committee, divisions were grouped into five institutes, each headed by a director whose role was managerial (CSIRO Annual Report 1977/78, p. 11). Divisional chiefs reported to institute directors, who in turn reported to the executive management council. In the decade following this review, a rise of economic rationalism was seen. *Economic rationalism* is an Australian term for a dogma that resembles what has been called *Thatcherism* in the United Kingdom and *Reaganomics* in the United States. This economic management approach, which has gained momentum, particularly in English-speaking Western countries, places a strong reliance on the workings of the free market.

## Effects of McKinsey Review

In this political-economic climate, the global management consulting firm McKinsey and Company was contracted to carry out a review of CSIRO in 1986. The acceptance of their recommendations resulted in the most far-reaching restructuring in CSIRO's history. Following from the McKinsey report, the executive management council, made up of senior CSIRO scientists, was abolished, and the organization was placed under the control of the CSIRO board, whose members were drawn largely from the business world. The five institutes established in 1979 following the Birch report were replaced with six institutes, each related to a section of the national economy. The previous 41 divisions were reduced to 32 divisions, and all units were amalgamated into existing divisions. It was further proposed that CSIRO should find more of its funds from non-Treasury sources. One of the recommendations adopted was that a vigorous program of management training for staff be introduced (CSIRO Annual Report 1986/87, p. 7).

I was attached to a unit whose mission was to carry out research to support the wheat industry, a large domestic and even larger export industry in Australia. The unit, located in Sydney, was making useful contributions to the industry. Among the contributions, methods had been developed for measuring protein and starch content and composition and for identifying wheat varieties. These advances facilitated the segregation of consignments for the different export markets, placing the country at an advantage with its competitors. The unit collaborated with an Australian company to develop an instrument for quantifying the quality of grain that had been subjected to varying degrees of weather damage. The uses of this instrument, the Rapid Visco Analyzer (RVA), have since been greatly expanded, and now it is used worldwide for a range of applications in the food industry. At the time, the unit was being awarded more than 50% of its funding by industry, and although small, its research was highly regarded internationally. The outcome of the McKinsey Report was that the unit became attached to a large division, whose headquarters were located in Canberra, a three-hour drive from Sydney. As invariably happens in a managerial climate, this division brought in a "new broom." The previous work of the unit was discredited, and a new approach was trumpeted. This new approach has not, however, contributed anything of value to the industry. The unit was later disbanded. Members were given the opportunity to move to Canberra under the guise that the work would be continued and enhanced there. Although some did relocate, there was not the level of expertise required to viably continue the work that had proceeded in Sydney, and so it fairly quickly petered out. I do not believe that McKinsey and Company bothered to find out anything about the work of the unit, or even knew it existed, so focused were they on the "big picture" involving management and restructuring. However, their involvement resulted in the demise of a resource that could have continued to make a useful contribution to one of Australia's foremost industries.

In a column entitled "The Moral Pygmies Who Run the Big End of Town," *Sydney Morning Herald* columnist Miranda Devine referred to an article by Malcolm Gladwell in *The New Yorker*. She wrote:

> Gladwell writes about how the famous management consulting firm, McKinsey & Company, drove a new management orthodoxy known as "the talent mind-set" across the corporate world. The best companies, they said, singled out and lavishly rewarded their "stars," recruited the best and brightest MBAs from the top business schools, pushed them into senior positions over their heads, prizing only their innate "talent and lack of experience." McKinsey spread the message "ardently" and of all its grateful clients, one was the most receptive. Gladwell writes "It was the company where McKinsey conducted 20 separate projects, where McKinsey billings topped $US 10 million a year, where McKinsey directors regularly attended board meetings and where the CEO himself was a former McKinsey partner. The company, of course, was Enron."*

Enron was an American energy company. Its share price hit a high of $US 90 per share in mid-2000 and then plummeted to less than $US 1 by the end of November 2001. It became the largest corporate bankruptcy in U.S. history up to that time, with consequent financial disaster for its shareholders and employees alike.

---

* Devine, Miranda, "The Moral Pygmies Who Run the Big End of Town," *Sydney Morning Herald*, July 21, 2002 (www.smh.com.au/articles/2002/07/20/1026898930296.html).

The "talent mind-set" promulgated by McKinsey has been evident in CSIRO. Those chosen as "stars" have had the ears of superiors and have used the privilege to tell them what they liked to hear rather than what they should have been told. Some of these anointed ones continued to be supported even after a history of blunders, while those who got on quietly doing their jobs were treated with disdain.

Other CSIRO laboratories experienced similar effects to what I have described above as a result of the imposition of managerial principles. For example, in 1988, the board decreed that the Division of Soils should aim to find 30% (later 25%) of its total funding from non-Treasury sources (Lee, 1998). As a result, it increasingly became necessary to design and work on projects that would be funded by sponsors prepared to fund short-term research. The planning and preparation of grant proposals and the corresponding reports that needed to be furnished occupied a great deal of time, mainly for the more senior scientists. Furthermore, the nature of grant-supported research is such that projects are problem oriented with short lifetimes, usually not more than three years. This lack of assured funding meant the erosion of long-term projects involving basic science. Previously, the division's capacity to support such research had enabled scientists to acquire the expertise that brought them recognition as leading authorities in their field, with the capacity to carry out research for the public good.

The changes introduced as a result of the McKinsey review represented a transition for CSIRO from a solidly based science organization dedicated to research to what was essentially a corporatized body with all the accompanying consequences, such as outsourcing and downsizing. Long-term research for the public good has tended to make way for short-term consultancies. In an Internet blog, John Quiggin gave a perceptive description of managerialism: "The main features of managerialism policy are incessant organizational restructuring, sharpening of incentives and expansion in the power and remuneration of senior managers, with a corresponding downgrading of the role of skilled workers and particularly professionals."* At the end of 1994, a *Sydney Morning Herald* editorial (cited by Rees, 1995) concluded, "So confused is the organization these days by the relentless probing and reorganizing that no one

---

* *John Quiggin Blog*, "Word for Wednesday: Managerialism" (www.johnquiggin.com/archives/001363.html), July 2, 2003.

has any clear ideas what its job is ... Politicians and bureaucrats love torturing the CSIRO."[*] In a blog, Shelley Gare, reported on how the number of highly paid management positions in CSIRO had risen sharply, stating that, "In the 2004–2005 annual report, there were five executives earning over $320,000. The next year there were nine and, in the 2006 report, there were twelve. Senior Research Scientists though are on between $100,000 and $120,000. In the six years to June 2004, while corporate positions were doubling, 316 people went from research projects."[†] The designation of the research and technical staff underwent a change during the period following the reorganization. Levels were assigned to different classifications. For example, senior research scientists became CSOF level 6, principal research scientists became CSOF level 7, and so on, an example of the downgrading of professionals referred to by John Quiggin.

One of the episodes of the BBC comedy program "Yes Minister" featured a hospital that had been operating for 15 months. It had 500 nonmedical staff and was being run very efficiently. However, there had been no patients or medical staff since its inception. Could it be that some scientific research organizations reach a stage where they are being run efficiently but will have no genuine motivated scientists left to do the research?

## Freedom of Expression in Science

Another effect of the encroachment of management on science has been the prevention of scientists from expressing their opinions on areas of their expertise. In an article in the *Canberra Times*, Rosslyn Beeby[‡] reported on a speech by a distinguished CSIRO scientist, Dr. Hugh Tyndale-Biscoe, on the occasion of the launch of a new edition of his influential textbook. Tyndale-Biscoe said that CSIRO had previously given primacy to scientists, not administrators—distinguished scientists led it, and the administration was dedicated to supporting the scientist and technician at all levels. He said that modern management practices were comparable to 1930s Soviet

---

[*] Editorial, *Sydney Morning Herald*, December 16, 1994 (as cited by Rees, 1995).
[†] *Contaminated Life Blog*, Robyn Williams, "Nothing Is the New Something" (Transcript of an interview with Shelley Gare) (liveness.org/contaminated-life/?p=54), February 25, 2007.
[‡] Beeby, Rosslyn, "Noted Ecologist Slams CSIRO," *Canberra Times*, May 7, 2005, p. 1.

science minister Nicolai Bukharin's determination to "neutralize" scientific opposition to government planning, and this was compromising scientific independence. A recent example of the attempted gagging of a scientific viewpoint, which was prominently featured by the media in late 2009, has been the treatment of CSIRO scientist Dr. Clive Spash. Spash submitted a paper to *New Political Economy* which was peer reviewed and accepted for publication. In the paper entitled "The Brave New World of Carbon Trading," it was suggested that emission trading schemes, which the government had hoped to introduce, were not the answer to climate change. CSIRO top management prevented Spash from proceeding to publish unless the manuscript was altered, something the author and the journal found unacceptable. In a bitter dispute, Spash, who had been head hunted by CSIRO, resigned, claiming to have been subjected to harassment and intimidation as well as censorship. Although the topic of leadership will be discussed in the next chapter, it is timely to refer to one distinguishing feature that relates to leadership. Good leaders are those who are loyal to their subordinates, providing they are doing their jobs as they are expected to. In contrast, poor leaders ingratiate themselves with those above them in the hierarchy and hang their subordinates out to dry, even though they may be carrying out their duties in the correct manner. The latter type of leadership (or lack thereof) is a characteristic of the managerial culture and not of the scientific culture.

Although not directly related to the chapter topic, it may be opportune to follow the reference to emission trading by referring to the controversial topic of climate change. There is strong opinion that there is a significant human contribution to warming of the planet. It is interesting to question whether this has the status of a scientific theory according to the criteria proposed by Karl Popper, discussed in Chapter 3. This is not to question whether the theory is correct or not but to examine how well it has been tested as a scientific theory. It is widely accepted that human effects on climate is a complex issue. Much evidence has accumulated which can be rationalized on the basis that emissions resulting from human activity are causing an increase in global temperature. But has any experiment been done with the aim of refuting this theory? It does not necessarily need to be a new experiment. Once a theory has been proposed, certain predictions logically follow. It is possible to test predictions by making observations on what has happened in the past. The question posed is whether any refutable tests have been made or whether all the observations that support the theory are simply confirmations of previous observations in the same manner as those of Adler's psychological theory (see Chapter 3). If not, it can only be considered to be similar to the myths or precursors of scientific theories as described by Popper. An interesting experiment that could serve to test the theory was outlined by Michael Asten in an article entitled "CSIRO Should Establish If There Was Medieval Warming Down-Under."* This relates to what has been referred to as the Medieval Warm Period (MWP) that is supposed to have spanned a range of time between the 9th and 15th centuries. There is some evidence in parts of the Northern Hemisphere that the rate of warming and temperatures attained were similar to what has been observed in recent decades. Asten has proposed to carry out a study of fossils, cave deposits, and tree-ring records from tropical to Antarctic Australia and territories to test whether the warming evidence was a global phenomenon. If this study would show a similar warming period to the MWP, then the basis of present belief that human contributions are causing today's warming would be undermined. If the experiments failed to show this, it could then be taken as a failed refutation, and the theory of human effects would be corroborated and emerge stronger. Asten suggested that this is an opportunity for CSIRO and the Australian Bureau of Meteorology to carry out the work. Based on the pronouncements of CSIRO top management to the effect that the theory has already been established, it seems doubtful if the suggestion would be taken up.

## A False Premise

The notion that business management principles can be applied to scientific research is based on a false premise. The premise that the performance of all organizations can be optimized by applying generic management theory is erroneous in the case of scientific research. In a transcript from *The World Today* entitled "CSIRO

---

* Asten, Michael, "CSIRO Should Establish If There Was Medical Warming Down-Under," *The Australian*, May 13, 2010 (www.theaustralian.com.au/news/.../ story-e6frg6zo-1225865724876).

Resembles Public Corporation: Leading Scientist," David Mark reported that former CSIRO scientist Hugh Tyndale-Biscoe said, "I emember some years ago when we were starting to be redirected in CSIRO and somebody said to me in the administration 'What will you be doing next year?' And I said 'Well, in the words of the Nobel Laureate, Szent-Gyorgi, if I knew what I was doing next year it wouldn't be worth doing because it would mean that it was pedestrian stuff.'"* This is the essence of the reason why scientific research cannot be run by business management principles.

## Performance Criteria for Scientists

The performance criteria for technical and research staff in some research organizations are formulated by administrators. At the beginning of the review period, staff members are required to document such things as objectives, tasks to be performed, time-lines, milestones, and expected outcomes. At the end of the review period, the expected outcomes and milestones are evaluated against the objectives. Although some flexibility may be allowed, this rather rigid framework conflicts with the way that scientific research should proceed according to a statement by Albert Szent-Gyorgi in regard to grant proposals. In an article entitled "Research Grants," Szent-Gyorgi (1974) wrote,

> Research means going into the unknown, which demands a pioneering spirit. This spirit is now strangled by the way in which the main federal biomedical granting agency, the U.S. National Institute of Health (NIH) distributes its grants. The unknown is unknown because one does not know what is there. If one knows what one will do and find in it, then it is not research any more and it is not worth doing. The NIH wants detailed projects, wants applicants to tell exactly what they will do and find during the tenure of their grants, which excludes unexpected discoveries on which progress depends. (p. 41)

---

* Mark, David, "CSIRO Resembles Public Corporation: Leading Scientist," *The World Today* (ABC Local Radio) (Transcript), July 4, 2005 (www.abc.net.au/worldtoday/content/2005/s1406264.htm).

The same reasoning applies to the performance reviews of research scientists. For any project, a researcher needs to begin with goals, but these need to be flexible so that when new insights into the problem are gained, the direction of the research can be easily changed. This is not possible with the rigid structure imposed by managerialism. As a result, scientists are often straightjacketed in their projects. When it comes to their performance reviews, they are locked into the tasks to be performed and the time lines and try to fit the milestones and conclusions to the expected outcomes in order to obtain a successful evaluation. Their creativity is thus stifled, and the research becomes pedestrian.

## Influence of Managerialism on Scientists

Confirmation of the inappropriateness of applying management ideology to scientific research comes from the loss of morale and job satisfaction and the steep increase in stress with its accompanying health problems suffered by CSIRO staff since the advent of managerial policies (Rees, 1995). The problems stem from an unhealthy work environment. Diane Cory (1998), in an article entitled "The Killing Fields: Institutions and the Death of Our Spirits," wrote the following:

> There is a lie that must be named and a truth that must be told. Our institutions are killing our spirits. We are allowing it to happen. In exchange for an illusion of power and control, safety and security, we have betrayed our souls because we are afraid. We are afraid to think for ourselves inside organizations. We are afraid of our bosses and our bosses' bosses. We are afraid of our colleagues and what they might say and how they might betray us in meetings or behind our backs. We are afraid we won't meet our deadlines. We are afraid of foreign competitors. We are afraid to "dream great dreams." We are afraid to be kind.

This, I believe, is an apt description of the working environment that has been foisted on CSIRO as a result of the imposition of managerialism.

In the mid-1990s, an active forum was established on the Web in which CSIRO staff contributed, mainly to air their dissatisfaction with the current trends. Nearly all the contributions had a consistent theme in which "top-down management," "command and control," and "emperors with no clothes" were recurrent phrases. The hope was expressed that when the new chief executive officer (CEO) was appointed, the present trend would be halted, and there would be a return to sane management. This was not to be. Contributions to the forum began to dry up. Some of the active contributors retired or resigned, others were retrenched, and some were taken out to lunch by management. Soon after, a colleague from another CSIRO division who had left before I did to take a university position, made an interesting observation. When he returned to visit his old division a short time later, he remarked that he felt that he had entered the "land of the living dead." The Killing Fields, so aptly described by Diane Cory, had taken their toll. Spirits had been crushed. People did not wish to risk unemployment by rocking the boat. Many had to support families, so this was their priority.

In my final years at CSIRO, I experienced firsthand the influence of managerialism on scientific research. Hierarchical structure was firmly embedded. Below the chief and assistant chiefs came the program leader, then a succession of subprogram leaders and project leaders. These were not appointed through open competition. They were not chosen because they were the best scientists, but were chosen more on the basis that they would fit well into a corporate system. It was no coincidence that many were characterized by having excessively large egos, which, I believe, is a predominantly male characteristic. Although designated as "leaders," in my opinion, many were managers who often used their status to subjugate subordinates and enhance personal ambitions. Of course, ego energy is necessary, as pointed out by Schuster (1998), because without it, a person would be ineffectual. Schuster suggests that a human person is a combination of ego and self. What is important is to have the right balance so as to avoid the extremes of an out-of-control ego that makes a person unbearable to work with or an ego so lacking as to reduce a person to a wimp.

Once a management culture is entrenched, its nature is such that it is difficult to bring about change. Organizations usually have a mixture of individuals who respond in different ways. Some see that things are wrong but are too apathetic or too afraid to act. Some are opportunists who curry favor with the managers because they know that by acting as lackeys, they will be rewarded, as happens in all totalitarian regimes. There are also those who act as double agents. Others would like to make changes but are powerless to do so. One or two try to change things, but because they are few, they are targeted by management and can easily be neutralized. Thus, there is usually little cohesion, and management finds it easy to divide and rule. Rees (1995) stated that "Managerial fundamentalism is apparent in its dogma, intolerance of critics, and gratitude for compliant staff" (p. 25). This is the exact opposite of what is inherent in the scientific method as discussed in Chapter 3. We thus have the paradox of a scientific organization under the control of a culture that is alien to science. It should be made clear that those who have accepted and applied the business management ideology to scientific research have not all been graduates from business management schools. Many who have helped to impose the ideology have been graduates from university science departments. We need to ask what is wrong with our university science departments that they are turning out people who do not seem to understand science and are content to contribute to its destruction.

## The Risks of Corrupt Practices

Another consequence of a rigid hierarchical system is the opportunities it opens up for corrupt practices. The top-down management structure is able to hide injustices. The powerful always protect the powerful. Some enlightened organizations have added bottom-up evaluations, facilitating identification of corrupt or incompetent managers by subordinates. However, in organizations where managerialism is entrenched, those who hold power are usually not prepared to risk the possibility of relinquishing any power by adopting this fairer system.

Rather than continuing with a theoretical discussion, I will recount my own experience in regard to promotion to illustrate the point. After repeated annual requests, I was finally given the go-ahead, and my case for promotion was put forward by the chief of the division. The case proceeded to the next higher (and normally the highest) level which was the institute. Following the meeting of institute directors to review promotion cases, I received a congratulatory letter from one of the referees of my case informing me that my institute director had sent him a letter thanking him for his input and informing him that my promotion had been recommended. At that time, CSIRO was undergoing a change in structure. A level of executive above the institutes had been put in place, and my former institute director had been elevated to the executive management council. I later received a letter from him (now in the capacity of executive member) informing me that my promotion had been denied. When I visited him to inquire about the denial, he told me that it was because of external advice. On pressing him about the external advice, I was told that it came from the chief of my division. In other words, my promotion case had been put forward by the chief, it had passed the institute level, been sent on to the executive council, and then been rejected as a result of external advice from the chief. Obviously, the other members of the executive council, who were entrusted to uphold the high ideals of the organization, had not been concerned about this abuse of due process.

In the preparation of my case, I put forward the names of four referees. When the division submitted my case, two of the names were used, and the division included one other. I was not consulted to see if I approved, although the regulations state that this is what should happen. This person was someone with whom the program leader had close ties. The report from this referee was highly complimentary but included one negative comment, namely that I had not shown leadership in the formation of a cooperative research center. The truth was that I had wished to take a leadership role but had been removed by the program leader. A single negative comment in a referee report often results in rejection of a case. Although it did not have this result at the institute level, it was to be made use of by my "opponents" when I lodged an appeal.

Prior to the appeal hearing, the program leader informed me that at least one of the appeal committee would not be on my side. I did not take much notice of this remark at the time, as I had not at that stage been informed of the composition of the committee. On the day of the appeal hearing, the chief, assistant chief, and program leader all traveled from Canberra to Sydney to participate, and each gave negative inputs. The appeal committee consisted of four members: the chairperson and three members. Because a senior promotion was being considered, one of the committee members had to hold the position of chief of a division. When the committee met after the hearing, they could not reach an agreement and so decided to submit two reports. One report, termed the *majority report*, signed by three members of the committee, stated that the requirements for promotion to the next level had been fulfilled and recommended the promotion. The other report, termed the minority report, signed by the divisional chief, recommended against promotion. I will leave it to the reader to connect the dots. I was later advised by the CEO that the appeal had been unsuccessful and that more weight had been given to the opinions of two chiefs (one a member of the appeal committee and the other the chief of my division). Neither of these chiefs had any understanding of my research, whereas the other three committee members did have some knowledge. Although I protested, the CEO advised, in a letter that arrived over two months later sent by surface mail, that he was not prepared to change his decision.

I have recounted my experiences to alert younger scientists to the dishonesty that can be found in high-level management and the contempt for due process that can occur in a system where managerialism operates. My advice in this climate is that one should always look for the good in people but with the qualification not to be naïve and to always be ready to deal with corrupt behavior. When I was preparing my promotion case, I was given access to previous cases at the same level that had been put forward in the division. One, in particular, attracted my attention. It was one in which the case was based on the introduction of "new genes" that would revolutionize the industry. No such genes existed; it was purely fictitious. Still, those who deliberated on the case would not know that, and it would have been a shame to permit this fact to spoil what was an excellent promotion case.

Schuster (1998), in an article entitled "Servants, Egos and Shoeshines: A World of Sacramental Possibility," paraphrases Roberto Assaglioli, a famous Italian psychologist and visionary, as asking the question, "Why is it that truly good people seem not to be powerful and the truly powerful seem not to be good" (p. 273). A simplistic answer to this question in regard to science may be that good scientists are dedicated to their profession and are not too interested in power. On the other hand, those who attain power are those who relentlessly seek it. I have observed that these people strive to always win. If winning can come by fair means, that is fine, but if not, other means will be resorted to. Winning is always the goal, whether it be to reach a position of power or to come off best in a personal conflict.

Managerial bosses think they know how to get to the bottom line based on outcomes that had been expected, milestones that had been predicted, and time lines that had been strictly adhered to. It is of little importance to some of them if this bottom line is achieved in a working environment characterized by demoralization, treachery, mistrust, and lack of humor. What if things had been different in the past few decades? What if CSIRO had been presided over by a government with an understanding of science and the vision of the one in 1949 when the organization was formed and by a minister with the statesman-like qualities of CSIRO's first minister, R.G. Casey? What if scientists had been allowed to use their creative talents to change direction and follow insights that could have culminated in unexpected discoveries of major importance? What if long-term research for the public good by scientists who had achieved international prestige had been strongly supported? What if this could have been done in an atmosphere of collegiality, kindness, and humor? What then would have been the bottom line? The managers would not be able to answer this, and none of us will ever know.

## Coping with Effects of Managerial Stress

No doubt, there are many research scientists in a situation in which they feel oppressed by those in their managerial hierarchy who are often not concerned with the welfare of their subordinates. In such a situation, the best advice is to leave and find another position. This, however, is usually not very practical advice. Research scientists are highly specialized. It is difficult to find another position that matches their expertise. Until one is found, it is necessary to find ways to best cope with the situation. Studies of the reactions of staff in different occupations have found that when a company is likely to shut down or employees are under the threat of retrenchment, those who cope best are those who continue to find meaning in their work. Unlike their colleagues, who resign themselves to their fate and sit around despondently waiting for the day of execution, those with a more positive attitude shut off a lot of what is going on around them and throw themselves into their work. In the case of scientists, this means continuing to focus

their minds on the problems being tackled, carrying out experiments and publishing papers. Even if they do lose their jobs in
the future, they will have utilized their time productively, which
will be beneficial for their careers. It may be that they will need
to apportion a certain percentage of their time to dealing with
problems thrust upon them by their managers. Sometimes, it can
be a bit like a chess game. It may require development of skills
in escaping from "check." The only difference is that unlike in a
chess game, it is only possible to defend, not to attack.

## References

Cory, D. 1998. The Killing Fields: Institutions and the Death of Our Spirits. In
*Insights on Leadership. Service, Stewardship, Spirit, and Servant Leadership*
(Larry C. Spears, ed.), John Wiley & Sons, New York, pp. 209–215.

Lee, K.E. 1998. A History of the CSIRO Division of Soils 1927–1997. CSIRO
Land and Water Adelaide Technical Report 43/98, November 1998.

Rees, S. 1995. The Fraud and the Fiction. In *The Human Costs of Managerialism.
Advocating the Recovery of Humanity* (S. Rees and G. Rodley, eds.), Pluto
Press, Sydney, Australia, pp. 15–27.

Schuster, J.P. 1998. Servants, Egos, and Shoeshines: A World of Sacramental
Possibility. In *Insights on Leadership. Service, Stewardship, Spirit, and
Servant Leadership* (Larry C. Spears, ed.), John Wiley & Sons, New
York, pp. 271–278.

Szent-Gyorgi, A. 1974. "Research Grants." *Perspectives in Biology and Medicine*
18:41–43.

# A BOAT RACE

*(www.wolfescape.com/Humour/Names.htm)*

Once upon a time it was resolved to have a boat race between a Japanese team and a team representing the N.H.S. Both teams practiced long and hard to reach peak performance. On the big day they were as ready as they could be.

The Japanese won by a mile.

Afterwards the N.H.S. team became very discouraged by the result and morale sagged. Senior management decided that the reason for the crushing defeat had to be found, and a working party was set up to investigate the problem and recommend appropriate action.

Their conclusion was that the Japanese team had eight people rowing and one person steering, whereas the N.H.S. team had eight people steering and one person rowing.

Senior management immediately hired a consultancy company to do a study on the team's structure. Millions of pounds and several months later they concluded that "too many people were steering and not enough rowing."

To prevent losing to the Japanese next year, the team structure was changed to three "Assistant Steering Managers," three "Steering Managers," one "Executive Steering Manager" and a "Director of Steering Services." A performance and appraisal system was set up to give the person rowing the boat more incentive to work harder.

The next year the Japanese won by two miles.

Following this, the N.H.S. laid off the rower for poor performance, sold off all the paddles, cancelled all capital investment for new equipment, and halted development of a new canoe. The money saved was used to fund higher than average pay awards to Senior Management.

## MCKINSEY REPORT ON A PERFORMANCE
## OF THE BERLIN PHILHARMONIC

*(www.year01.com/forum/issue1.htm)*

The four oboe players do not have much to do for quite a long period. This part should be shortened and the work should be equally distributed on all members of the orchestra in order to avoid peak loads. The twelve violins play all the same melody. That is an unnecessary parallel work. This group should be made drastically smaller. In case a higher volume of sound is preferred this can be achieved also by an electronic amplifier. To play parts which include 1/32 notes requires a large work effort. It is recommended to condense all these notes to 1/16 notes. Then even students of music and less qualified musicians could be employed. In some passages there is too much repetition. The full score should in this respect be thoroughly revised—what is the use of it if the horn repeats a passage which has been played already by the violins?

If all those unnecessary passages are eliminated then the concert, which takes up to two hours in time, will last approximately twenty minutes. This implies that the intermission can be omitted. If all these measures do reduce demand, then part of the concert hall could be closed which saves money for personnel, electricity, heating, and so forth.

# CHAPTER 6

# Leadership in Science

If we are to recover from the negative influences that managerialism has had on scientific research, we need leaders to show the way back. The term *leader* conjures up different images for different people. These range from ruthless tyrants like Hitler, Stalin, and Pol Pot to the opposite end of the spectrum where we find the servant leader exemplified by Robert Greenleaf. Diane Cory (1998), in an article entitled "The Killing Fields: Institutions and the Death of Our Spirits" (mentioned in Chapter 5), writes "I believe that by living in fear and accepting it as the norm in our institutions, we have confused positions of authority with leaders and abuse of power with leadership" (p. 212). Everyone in our society can be a leader. In the scientific community, some of those who have assumed positions of authority are not leaders, they are often just "powertrippers." On the other hand, some who have no position of authority are leaders, setting an example to others by their integrity and honesty toward their work and in dealings with colleagues. Leadership is not something that can be easily defined, but usually it is possible to recognize when it is present and when it is not.

During one of the many restructurings in the Commonwealth Scientific and Industrial Research Organization (CSIRO) that I referred to in the previous chapter, two divisions were to be amalgamated. The chiefs of these two divisions had a difference of opinion on how this should be done and went to see the chief executive officer (CEO). The CEO promptly told them to go away and come back once they had agreed on the new structure. A colleague of mine remarked that this was an example of leadership. Evidently what had impressed my colleague was the decisiveness of the CEO. This attribute of solving problems by making quick and firm decisions is what is regarded by many as strong leadership. For me, this example was the antithesis of leadership. A good leader would have engaged the two chiefs, initially separately, to understand the position from where each was coming. Then, he or she would have tried to negotiate a compromise that would be acceptable to both so as to result in a win-win situation. To do this requires certain skills, but, above all, this requires being able to listen receptively and to empathize with others. In environments where managerial tactics operate, these attributes are not easily found.

## Mentoring

At some stage in their careers, scientists will almost certainly be expected to act as mentors for less-experienced colleagues. This may mean interacting with colleagues in the same or in a different institute. It may involve someone seeking advice in regard to interpretation of their data or on how to proceed with his or her project. Most scientists have a generous spirit and are happy to pass on the benefits of their experience, finding this rewarding. More formal mentoring involves participating in an organized mentor/protégé program run by an organization or serving as a major advisor for a graduate research student. It is in this latter capacity as advisor to a budding researcher that I would like to pass on some thoughts. Of course, there are no rules for this, and the approach of a supervisor may vary greatly depending on the individual and also on the mentee. Later I will say a little more about the role of a leader, but first I would suggest that the most essential quality is integrity. The student needs to have confidence that the advisor will always be honest and will always be supportive. I have heard it said that where a student behaves badly such as by being uncooperative, dishonest, or just lazy, the supervisor should keep a written record of the student's transgressions, presumably so as to use it when it comes to a final showdown. It is a bit like keeping a record of your spouse's bad behavior so that you will have it ready when the divorce proceedings begin. This is completely unacceptable. It is essential that there be a relationship of complete trust between advisor and student. Even if a student errs in behavior, the supervisor must never betray the trust.

## Supervision of Research Students

In a university science department, a faculty member will usually be in charge of a laboratory and will be expected to act as major advisor for a number of graduate research students. This position carries certain responsibilities. The supervisor has the responsibility of ensuring that the students are provided with the conditions required to carry out their work, culminating in successful completion of their postgraduate degree. It is the duty of the supervisor to ensure that the student receives sufficient remuneration so as not to experience economic hardship. This may entail seeking external funding to support the research. The student should be provided with a viable research project and the laboratory facilities and equipment needed to perform the work. Next, the supervisor should endeavor to create an environment conducive to research. This includes stimulating the students and encouraging harmonious relations between the different members of the group. This can be helped by organizing small social functions in the laboratory, such as lunches or teas. Short, informal meetings should be held regularly for students to raise any problems they may have encountered such as equipment that is not working. It can also be an opportunity to give short summaries of their research projects and the obstacles they are encountering, with the goal of getting feedback from their fellow students. All this helps to foster collegiality. The importance of the choice of a viable research area cannot be overemphasized. Students have a finite time to complete the research, so it can be demoralizing to have to spend long periods banging their heads against brick walls. To begin with, the supervisor usually will need to suggest the problem and give some ideas how to tackle it. As time goes on, students may take an active role and, ideally, will proceed to formulate their own original approaches.

The way research students are supervised varies greatly from one supervisor to another. Some supervisors place goals for their students and then coerce them to achieve these goals. This resembles somewhat the managerial system described in the previous chapter. Rather than giving an opinion as to what is the best way to oversee students, I will briefly describe the way I have tried to guide students and the reasons for adopting the approach. I believe that research is an activity that requires a certain amount of tranquility, free from too much external pressure. I have been criticized because I did not supervise students closely enough. I understand how this perception may have arisen, but I think it is one that fails to take into account the true nature of research. I heard once through the grapevine that one of my students, who was not happy with progress, had said that I did not push my students. This was quite true.

I believe that an essential attribute that a researcher must develop in his or her training is self-motivation. Once someone becomes self-motivated, the drive will then come from within and not from outside, and it will be very much more effective. If someone carrying out research feels that he or she has not progressed as much or as quickly as expected, it is illogical to seek to attach the blame to someone else. Persons who choose scientific research as a career and do not become self-motivated may need to rethink their vocation. Self-motivation in research means developing a burning passion to tackle the unknown to the point that it is almost an obsession. Of course, this is meant in a positive sense and does not mean adopting any unhealthy aspects of obsessive behavior.

There are no rules for the regularity of discussions between advisor and student. It will vary. Some students like to have frequent discussions. Others prefer to have a good deal of freedom to follow their leads and have something tangible before talking to their advisor about their progress. The advisor must always be vigilant to ensure that a long period between meetings does not mean that the student is baffled and does not know how to continue. I believe that the approach of an advisor to discussions with students should be similar to that adopted by a psychoanalyst. The importance of research students being able to arrive at their own insights into a problem is often not appreciated. When students present results and are unsure how to interpret them and how to proceed, a common mistake is for the more experienced advisor to jump in and provide explanations. This prevents students from developing their own ideas. When the advisor sees the solution to a problem and does not convey it, this gives the student an opportunity to arrive at the solution. This may happen quickly or may take some time. Patience and discipline are therefore required on the part of the advisor. If this practice is followed, it becomes second nature for the advisor. This does not mean that the advisor should deliberately withhold information. It means that the advisor may point the student in the right direction by suggesting a paper or a section from a textbook, without explicitly providing the answer. In this way, it allows students to arrive at a clearer appreciation of the problem and gives them a sense of achievement and the confidence to tackle future problems. There have been occasions when students have come to discuss their research and, at the end of the session, have become animated and thanked me profusely for helping their understanding. I have only sat and listened without contributing anything. By talking about the subject, they have been able to clarify their thoughts.

When I retired from the university, two of my former graduate students traveled to see me. They both had recently married and brought their wives. They kindly invited me and my wife for a dinner. During reminiscences, they remarked that often before students went to meet with me, they would be agitated and worried, but after the meeting, they would appear serene and happy. This was a complete surprise to me as I had never deliberately set out with that purpose. Regardless of whether it was true or not, I think most advisors would wish to be able to have this effect on their students.

## Qualities Needed to Lead Scientific Research

In the previous chapter, I discussed a few of the trends in leadership of a scientific organization which have resulted from the adoption of ideology based primarily on economic outcomes. This has produced demoralization, stress, and accompanying health problems for scientists. It is obvious to many that there needs to be a reversal of these trends by restoring human well-being. Arguments as to why the managerial culture is not appropriate for scientific inquiry to flourish were put forward in the previous chapter. Suggestions as to how this may be accomplished for organizations in general have been outlined in the book edited by Rees and Rodley (1995). Some of the qualities needed to lead scientific research are as follows:

1. Broad vision and an understanding of science, including its philosophy and history
2. Recognition that the greatest asset in research is not the million dollar grant or instrument but the individual scientists, because it is their creativity that produces original ideas that can lead to advances in knowledge
3. Concern for the well-being of scientists in their charge, meaning that they should be treated fairly and equally without the cronyism and favoritism that is sometimes evident where the ruthless managerial style operates; a good scientific leader works to enhance the careers of subordinates, whereas poor leaders trample on their best interests
4. Minimization of hierarchy, but where it exists, the most gifted scientists should be given the leadership roles

## Servant Leadership

The third of the above qualities emphasizes leaders' concerns for their subordinates, an attribute that leads naturally to a concept that has achieved prominence in recent decades, that of servant leadership. This was introduced by Robert Greenleaf and described in several publications (Greenleaf and Spears, 2002; Frick, 2004). Since his passing on in 1990, others have continued his pioneering work at the Greenleaf Center for Servant-Leadership in Indianapolis, Indiana. The ideas of servant leadership have been disseminated by others in the form of articles and conference presentations and have been introduced into some organizations. Briefly, the concept is that the servant leader is a servant first but later chooses to aspire to leadership. Greenleaf received inspiration from reading Herman Hesse's *The Journey to the East* (1956). In this story, a group of men set out on a spiritual journey sponsored by an Order. One of the group, Leo, was the servant who sustained other members of the group by his spirit. The journey was going well until Leo disappeared. The group then fell into disarray, and the journey was abandoned. Later, the narrator of the story, one of the group, after wandering for years, is taken into the Order that had sponsored the journey. He then met up again with Leo whom he had known as a servant but now discovered that he was the guiding spirit of the Order and its noble leader.

The essential feature of servant-leadership is that the leaders' experiences as servants fit them to understand how to serve others. A good servant tries to insure that other peoples' needs are being served. In science, it follows that the leader gives priority to the best interests of the scientists. This means trying to insure the conditions for each scientist's creative talent to flourish and to remove obstacles to this objective. According to Greenleaf, the best test of servant leadership is "Do those served grow as persons? Do they, while being served, become healthier, wiser, freer, more autonomous, more likely themselves to become servants?" In regard to the future, Greenleaf (1998) believed that

A new moral principle is emerging, which holds that the only authority deserving one's allegiance is that which is freely and knowingly granted by the led to the leader in response to, and in proportion to, the clearly evident servant stature of the leader. Those who choose to follow this principle will not casually accept the authority of existing institutions. Rather, they will freely respond only to individuals who are chosen as leaders because they are proven and trusted as servants. To the extent that this principle prevails in the future, the only viable institutions will be those that are predominantly servant-led. (p. 17)

Of course, these ideas will be vigorously opposed by those who believe in coercive power and control rather than a trust based on mutual accord. Not everyone can fit into a servant-leader culture, and this was admitted by Greenleaf. In his biographical notes (cited by Frick, 1998), he refers to his foreman telling a young worker, "You know, if a fellow is an S.O.B., deep down inside, he had just better go ahead and be one, because if he tries to be something else, he will likely be seen as both a hypocrite and an S.O.B., and that's worse" (pp. 354–355).

## The Inverted Pyramid

The concept of servant leadership leads to a structure that is, to some extent, opposite to that found in most organizations. The usual structure for organizations is a pyramid in which the CEO is at the apex, and successive layers of management move downwards. When it comes to matters of vision, mission, values, and setting of major goals, it is essential that the pyramid remains upright (Blanshard, 1998). However, Blanshard suggests that the implementation of policies can often be made in a better way if the pyramid is inverted. When this happens, those at the top of the organization become those who interface with the customers. Those at the bottom of the pyramid are now the top managers. Their role is then to support those at the customer interface and help them to accomplish their goals. These goals are generally to provide excellent service to the customers. We have all experienced the frustration of dealing with bureaucracy when we have been unable to have our reasonable requests satisfied because the people we deal with are far removed from the top management and have no autonomy.

Let me relate one of my experiences in this regard. A colleague from a scientific institute in Mexico arranged for her graduate student to spend two months at my laboratory in the United States to gain experience in an experimental technique. Although the immigration department had given assurance that a visa would be issued to the student in time, there was an unexplainable delay in providing it. The result was that the student's visit was delayed, and she was only able to spend two weeks in the United States. In order to have the required equipment and reagents for the student to work with, the Mexican institute requested that we arrange for their purchase through the U.S. university and they would reimburse us, which they did. This was done to avoid the expected long delay in acquiring the goods. The student, however, did not have the time to use the materials and was only able to gain some preliminary training. After the student returned home, we attempted to forward the materials to the Mexican institute, which was fair as they had paid for them. The university administrator sent me a two-page e-mail explaining why it was not possible to send goods that had been purchased through the university to another institute. I took the matter to my department head who solved the problem in an instant and authorized dispatch of the goods. They were sent off by FedEx after complying with all the paperwork. On their arrival in Mexico, my colleague went to the airport to pick them up but was told that she would be unable to do so as there were two reagents that were on the list of hazardous chemicals. We scientists used these chemicals regularly, knew they were listed as hazardous, and were completely familiar with the precautions for using them. This example shows how, at every step, guidelines were followed blindly without considering what was just and in the best interests of those who were being served. Those providing service continued to draw their regular salaries for "doing their job." In contrast, those supposedly being served, were all disadvantaged—the student by a lost opportunity and the two science departments by a waste of their precious resources. By inverting the pyramid, the employees at the interface with customers are now responsible for implementation of policies. They then go out of their way to be of service. The role of the managers, who are now lower in the inverted pyramid, is to give support to the employees and insure that they are given the resources to serve the needs of the customers. Under these conditions, common sense prevails, and frustration is replaced by satisfaction for the customer.

Although the example given was for business administration, the same principles can be applied to a scientific organization. The role of the leaders is to give support to the scientists and insure that they have all the resources and encouragement to use their creative talents to pursue research. The satisfied customer becomes an appreciative public whose lives are enhanced by the advances made by science.

## The Future

Will leadership in scientific research change in the future and, if so, how? With the wisdom of hindsight, it is easy to understand how the present organization of many scientific institutions has evolved. Research is largely dependent on funding, as most research, especially that of a long-term nature, does not generate its own funding; although, if successful, its costs may eventually be recouped many times over. Most funding is provided by governments and industrial companies. Governments are accountable to taxpayers, and companies are accountable to shareholders. Each of these entities has the responsibility to allocate the funding to give maximum benefit. This is indisputable. What are in dispute are the conditions placed on scientists, purportedly to achieve the research goals. For example, government representatives primarily come from professions other than science and therefore do not have a deep understanding of science. The promises of managerial ideology are appealing. Greater efficiency and doing more with less seem to be plausible goals. If it comes with demoralization and an unpleasant work environment, it might be argued that this is the cost of economic success. But is economic success really achieved under these circumstances? If the creativity of scientists is curtailed, can optimum economic success result? If not, then the argument that an unpleasant working environment is the cost, breaks down. The potential advantages of scientific research in a freer environment are not quantified, and managerial principles have no way of estimating it.

It is becoming increasingly recognized that the way scientific institutions are run cannot be based solely on economics, although this always has to be an important component. Another important consideration is that scientists must be given the opportunity to develop their full potential. Thus, there has to be an appropriate balance between strategies to give economic success and ones that provide a workplace conducive to the well-being of the employees. In the case of scientists, the work environment needs to emphasize empowerment of individuals and be characterized by purpose and dignity. Younger scientists entering the profession or in the early stages of their careers inherit the challenge to help create this environment. As pointed out by John Renesch (1994), companies cannot change long-established cultures unless their leaders also change. Renesch (1994) states that "The modern business books are full of attempts to transform organizations without any material change by their chief executives or other senior people. The new leaders will be people of vision, who inspire others to become part of their vision. They won't convince or manipulate to recruit other people to join them; they will attract them, like a magnet" (pp. 2–3). Robert Greenleaf has also given suggestions on how change must begin by changing oneself if we are to move to organizations with high ideals. Service to others is at the heart of the servant-leader concept. However, to get the kind of trust that enables an empowerment culture to thrive, it is necessary not only to have individuals who are trustworthy and whose vision is shared with the organization, but to have a trustworthy organization that fosters and supports empowerment.

# References

Blanshard, K. 1998. Servant Leadership Revisited. In *Insights on Leadership* (Larry C. Spears, ed.), John Wiley & Sons, New York, pp. 21–28.

Corey, D. 1998. The Killing Fields: Institutions and the Death of Our Spirits. In *Insights on Leadership* (Larry C. Spears, ed.), John Wiley & Sons, New York, pp. 209–215.

Frick, D.M. 1998. Understanding Robert K. Greenleaf and Servant Leadership. In *Insights on Leadership* (Larry Spears, ed.), John Wiley & Sons, New York, pp. 353–358.

Frick, D.M. 2004. *Robert Greenleaf: A Life of Servant-Leadership*, Berrett-Koehler, San Francisco, California.

Greenleaf, R.K. 1998. Servant-Leadership. In *Insights on Leadership* (Larry C. Spears, ed.), John Wiley & Sons, New York, pp. 15–20.

Greenleaf, R., and Spears, L.C. 2002. *Servant-Leadership: A Journey into the Nature of Legitimate Power and Greatness/Essays by Robert K. Greenleaf*, edited by Larry C. Spears, Paulist Press, Mahwah, New Jersey.

Hesse, H. 1956. *The Journey to the East*. (Translated from German by Hilda Rosner.) Picador, New York.

Rees, S., and Rodley, G. 1995. *The Human Costs of Managerialism—Advocating the Recovery of Humanity*, Chapter V, Pluto Press, Sydney, Australia.

Renesch, J. 1994. A Commitment to a Change in Context. In *Leadership in a New Era: Visionary Approaches to the Biggest Crisis of Our Time* (J. Renesch, ed.), New Leaders Press, Sterling and Stone, San Francisco, California, pp. 1–5.

# CHAPTER 7

# Insights from Notable Scientists

In previous chapters, we looked at the philosophy of the scientific method and some of the attributes needed by research scientists for a successful career. In this chapter, we will focus attention on individual scientists who have attained eminence and try to obtain some insights into their success. Thus, the main aim will not be to attempt to give biographical sketches but to focus on some of the unique features of their careers that might hold the secrets to their achievements. It is in no way meant to be a list of the greatest scientists but will select primarily some of those who fit the objective. When assessing the accomplishments of scientists, it is important to take into account the era in which they worked and thus the extent of knowledge at that time. Thus, it is unfair to play down the achievements of those who worked at earlier times based on the understandings of today. Science progresses through building on the advances made by those who have gone before. As Isaac Newton remarked, "If I have seen further than others, it is because I have stood on the shoulders of giants."*

## Marie Curie (1867–1934)

Marie Curie (née Skodlowska) grew up in Poland. Her father taught mathematics and physics, and her grandfather had been a teacher in Lublin. Her family, both on her paternal and maternal sides, lost their property and fortune through patriotic involvement in Polish national uprisings. She went to Paris in 1890 but, initially, could not afford university tuition. She returned to Warsaw in 1891, tutored for a short time, and then returned to Paris at the invitation of her sister. After briefly staying with her sister, she rented a primitive room and proceeded with her studies in physics, chemistry, and

---

* Millard Fillmore's Bathtub (timpanogos.wordpress.com/.../quote-of-the-moment-newton-giants), accessed September 2010.

mathematics at the University of Paris (Sorbonne). She studied during the day and tutored in the evenings, scarcely able to earn her keep. She obtained a degree at the Sorbonne in mathematics in 1894. She returned to Poland, believing she would be able to work there in her chosen field of study. However, she was denied a place at Krakow University, apparently because she was a woman, and returned to Paris.

Her career was influenced by Becquerel's discovery in 1896 that uranium emitted rays that did not depend on an external source of energy but seemed to arise spontaneously from the uranium itself. She looked into this phenomenon as a potential field of research for a thesis. For the study, she used an electrometer, an instrument that had been invented by her new husband, Pierre Curie, and his brother 15 years earlier. She showed that the radioactivity of uranium compounds depended only on the quantity of uranium present, and that the radiation came from the uranium atoms, an important advance at that time. Uranium occurred naturally in two minerals: pitchblende and torbernite (also known as chalcolite). It was found that pitchblende was about four times as radioactive as uranium, and torbernite was twice as radioactive. It was deduced from this that these two minerals must contain small quantities of some other substance that was more active than uranium. Pierre Curie joined his wife in the work, and together, they announced in 1898 the existence of an element that they named *polonium* (in honor of her native Poland). In the same year, they announced the existence of a second element that they named *radium*, for its intense radioactivity.

Pitchblende is a complex mineral, and the chemical separation of its constituents was a hugely arduous task. Polonium was relatively easy, but radium was more difficult. Chemically, it resembled the element barium, and these two elements were both present in pitchblende. The Curies undertook the daunting task of separating out radium salts by differential crystallization. The problem was that the concentration of radium in pitchblende was much lower than had been anticipated. From a ton of pitchblende, one tenth of a gram of radium chloride was separated in 1902. By 1910, Marie Curie isolated the pure radium metal. In the meantime, her husband had been killed in an unfortunate street accident in 1906. In 1903, the Royal Swedish Academy of Science awarded Pierre Curie, Marie Curie, and Henri Becquerel the Nobel Prize in physics for their research on radiation. In 1911, Marie Curie was awarded the

Nobel Prize in chemistry for discovery of the elements radium and polonium, the isolation of radium, and the study of its nature and compounds. Thus, she became the first person to have been awarded Nobel Prizes in two different sciences. She also had the honor of becoming the first female professor at the University of Paris.

## Charles Darwin (1809–1882)

Charles Darwin was born at Shrewsbury, England, the son of a doctor. He began to study medicine at Edinburgh but did not pursue a medical career. He transferred to Cambridge to train for the ministry. At Cambridge, he befriended John Henslow, a biology professor, and Adam Sedgwick, a geology professor, and was motivated to become interested in biology and geology. Through the influence of Henslow, he was invited to accompany an expedition on the HMS *Beagle*, a ship organized by the Admiralty to chart the waters of the South Seas, under a young captain, Robert Fitzroy. The cruise of the *Beagle* extended to almost five years (1831 to 1836). The good facilities on the ship and the time spent exploring new lands gave Darwin a wonderful opportunity to make observations and to collect specimens, which formed the basis for the theories he was to develop over many years after the return of the *Beagle* to England. During the expedition, Darwin spent much of the time on land, while the *Beagle* chartered the coasts of South America, surrounding islands, and as far as Australia.

During the voyage of the *Beagle*, Darwin read avidly from the books in the ship's library as well as those he brought along himself. One book, Charles Lyell's *Principles of Geology* (1830–1833), influenced his thinking. Lyell proposed that the earth's geology had changed gradually over long periods of time as a result of cumulative local events such as earthquakes, volcanic eruptions, erosion, and deposits of sediments. This differed from the currently held view that changes occurred through violent, short-lived events such as the raising of mountains. Darwin's observations tended to confirm Lyell's views. For example, in Sao Tigre, a volcanic island about 1800 miles southwest of the Canary Islands, he deduced from the sequence of sediments that the island's surface had been formed by a succession of volcanic events, interspersed with subsidence and building up over a long time, and not a single violent one. He did, however, disagree with Lyell's opinion that coral reefs were formed by volcanic action. He believed that they were formed by gradual changes resulting from subsidence and elevation events. He predicted that if an island sank below the ocean surface, corals would continue growing, and a reef would turn into an atoll. Lyell accepted Darwin's interpretation, and its validity has since been confirmed by subsequent drilling.

His geological work brought him recognition in scientific circles, but the problem that particularly occupied Darwin's attention was the question of how different species of organisms originated. During his voyage on the *Beagle*, he was intrigued by many observations. For example, in South America, he had found fossils of armadillos that were now extinct but were similar but not identical to those presently existing. In Argentina, the giant ostriches (rhias) on the plains were different than the smaller species found in Patagonia, and both were different than the African ostrich. Thus, in these two examples, species were seen to vary as a result of both time and geographical location changes. The prevailing view was that as one species disappeared, a new species suddenly came into existence to replace it. Darwin found this unacceptable, and his thinking turned to transmutation (i.e., the idea that species changed from one location to another or from one era to the next by gradual changes over long periods that may be thousands or millions of years). The challenging question Darwin pondered was by what mechanism this was achieved.

After his return from the *Beagle* voyage, he had sent the specimens he had collected to experts in Cambridge and London for analysis. He worked on the information he got back together with the notes he had compiled during his voyage and concentrated on finding a mechanism for transmutation of species. At this time, he was influenced by reading Thomas Malthus' "An Essay on the Principles of Population" (1978). In this essay, Malthus proposed that population growth would be decreased by limits on food supply. Darwin realized that in the struggle for existence, favorable variations would tend to be preserved and unfavorable ones destroyed. The result over time would be the formation of new species. Thus, an individual with enhanced characteristics such as a sharper beak or a brighter color might be better able to survive and reproduce than others. This essentially was the Principle of Natural Selection. It shifted attention from differences between species to competition within species. Darwin resisted publication of his ideas for many years, during which time he consulted extensively with plant and animal breeders and thought deeply about the subject. Finally, he was hurried into revealing his theory by receiving a communication from another naturalist, Alfred Russel Wallace, in which he described a theory similar to that which Darwin had been elaborating. Wallace, however, did not have the large amount of confirming experimental evidence that Darwin had accumulated. It was agreed that Darwin and Wallace present a joint paper to be read to the Linnean Society of London in 1858. Darwin then went on to publish his theory in 1859 in a publication entitled "On the Origin of Species by Means of Natural Selection, or the Preservation of Favoured Races in the Struggle for Life" (1859). This has proved to be one of the most influential scientific papers ever written. Darwin's work, in simple terms, showed that evolution occurred by changes requiring thousands or millions of years. The mechanism for these evolutionary changes was called natural selection, a process in which species survived through the inheritance of favorable traits. Furthermore, the millions of species present today have evolved from common ancestors by a branching process called *speciation*.

## Albert Einstein (1879–1955)

Albert Einstein gained a diploma in physics and mathematics from the Swiss Federal Polytechnic School in Zurich. His immediate career was a little unusual. Unable to find a teaching post, he took a position as technical assistant in the Swiss Patent Office in 1901. It was during his time here and in his spare time that he produced some of his most remarkable work. The work at the patent office did not require great mental effort, and it is thought that this freed him to apply his intellect to important problems that attracted his interest. In 1905, he published three papers, each one a milestone in science:

1. The theory of special relativity that introduced revolutionary concepts about the physics of the universe. Up to that time, Euclidean geometry was used to describe space, using three spatial coordinates. Einstein showed that when large distances and high velocities (of the order of the velocity of light) were considered, it was necessary to bring in the extra coordinate of time, so that problems needed to invoke a four-dimensional space–time continuum. From the theory of relativity, one prediction was that a moving clock runs

more slowly and is shortened in length in the direction of motion, a result that seemed counterintuitive to previous thinking. This prediction has been confirmed by various experiments. Another outcome was the simple equation $E = mc^2$ ($E$ is energy, $m$ is mass, and $c$ is the velocity of light), which showed that mass and energy are interconvertible, the basic equation for describing nuclear reactions.

2. A theory of Brownian motion. This is a phenomenon in which particles suspended in a liquid show continuous irregular movements. The movements are caused by molecular collisions with the particles. Einstein developed a theory in which movements of the particles could be predicted from a simple equation involving the diffusion coefficient and time.

3. The photoelectric effect in which electrons are emitted from a surface, usually a metal, in response to incident light. This illustrated the dual nature of light (i.e., it could be thought of as waves or particles). Electrons were emitted provided the energy of the light was above a threshold value (a quantum). Einstein's conceptual work initiated discoveries that developed the quantum theory and formed the basis for modern electronics. The explanation of the photoelectric effect and other contributions led to the award of the Nobel Prize in 1921.

From 1911 to 1933, Einstein occupied professorships at Zurich, Prague, and Berlin. In 1933, he moved to the United States where he took the position of Professor of Theoretical Physics at Princeton University.

*The important thing is not to stop questioning.** (Albert Einstein)

*I never teach my pupils. I only try to provide the conditions in which they can learn.*† (Albert Einstein)

## Rosalind Franklin (1920–1958)

In 1938, Rosalind Franklin attended Newnham College, one of two women's colleges at Cambridge University, where she majored in physical chemistry and graduated with a bachelor of arts in 1941.

---

* The Quotations Page (www.quotationspage.com/quote/9316.html), accessed September 2010.

† The Quotations Page (www.quotationspage.com/quote/40486.html), accessed September 2010.

Following this, she started work with the British Coal Utilization Research Association (BCURA), where she was able to pursue research for her Ph.D. For the next four years, she worked on the elucidation of the microstructures of various coals and carbons. The aim was to explain why some were permeable to water, gas, and solvents and how heating and carbonization affected permeability. She demonstrated that pores in coal have constrictions at a molecular level, acting as "molecular sieves," successively blocking the penetration of substances according to molecular size. This fundamental work enabled the classification of coals and the accurate prediction of their performance to high accuracy. The research yielded five publications, and she was awarded her Ph.D. in 1945.

At the end of World War II, she took a position at the Laboratoire Central des Services Chimique de L'Etat in Paris. Here she learned the technique of X-ray diffraction and used it to analyze the structures of graphitizing and nongraphitizing carbons. This work helped in the development of carbon fibers and heat-resistant materials and earned her an international reputation among coal researchers. Although happy in France, she decided to return to England in 1949 and was awarded a three-year fellowship to work under the direction of John Randall at King's College, London, where she was assigned to work on the structure of DNA. The application of X-ray crystallography to elucidate the structure of complex biological molecules was just beginning, and Franklin was a pioneer in this area. Unfortunately, her stay at King's College was not a happy one due to a misunderstanding with a colleague, Maurice Wilkins, who assumed that she was to work as his assistant on the DNA research, whereas Randall had given her autonomy in her project. The climate for women at the university was not ideal at the time. For example, only males were allowed to eat lunch in the common room, and after hours, Franklin's colleagues went to men-only pubs.

Franklin discovered that DNA could crystallize in two different forms: the A and B forms. These two forms were mixed together, yielding impure crystals and making the X-ray diffraction patterns impossible to interpret. Franklin developed an ingenious albeit laborious method to separate the two forms, enabling clear X-ray patterns to be obtained. One of the patterns was shown by Wilkins to James Watson and Francis Crick, also working on the structure of DNA, at the Cavendish laboratory in Cambridge, without the authorization of Franklin. This helped Watson and Crick in their formulation of the structure of DNA, which they published in 1953, without properly acknowledging their debt to Franklin's work.

Franklin arranged to transfer her fellowship to J.D. Bernal's crystallography laboratory at Birkbeck College, London, where she turned her attention to the structure of plant viruses. Bernal, an eminent scientist, had called her X-ray photographs of DNA "The most beautiful X-ray photographs of any substance ever taken."* From 1956, Franklin became seriously ill and died in 1958, tragically cutting short a brilliant career. In her 16-year career, she published 19 articles on coals and carbon, 5 on DNA, and 21 on viruses. During her final years, she received increasing numbers of invitations to speak at conferences all over the world. Had she lived to continue her work on viruses, this would almost certainly have earned her further awards and professional recognition.

## Galileo Galilei (1564–1642)

Galileo Galilei was born in Pisa, Italy, but his family moved to Florence in the early 1570s. He matriculated at the University of Pisa, where he was to study medicine but became more interested in mathematics. In 1585, he left the university without having obtained a degree. For several years, he gave tuition in mathematical subjects. It was during this period that he designed a new form of hydrostatic balance and began studies on motion. His work on center of gravity brought him recognition among mathematicians, and in 1589, he was appointed to the chair of mathematics at the University of Pisa. A famous experiment that he carried out at that time was to drop objects of different weights from the Leaning Tower. This experiment showed that the time taken to fall was not proportional to the weight as Aristotle had claimed.

Later, after obtaining a chair of mathematics at the University of Padua (1592–1610), Galileo determined that the distance fallen by a body is proportional to the square of the elapsed time (the law of falling bodies). Also, he showed that the trajectory of a projectile is a parabola, again contradicting Aristotelian physics. It needs to be recognized, however, that despite these contradictions, Aristotle (384 BC–322 BC) was one of the giants who has influenced thinking.

---

* San Diego Supercomputer Center (www.sdsc.edu/ScienceWomen/franklin.html), accessed October 2010.

In 1609, there was to be a change that dramatically affected Galileo's career as well as subsequent science. An instrument (an early form of telescope) had been invented in the Netherlands that magnified images of distant objects. Galileo, by learning the art of lens grinding, was able to improve the instrument and to use it for study of the skies. Important discoveries followed with the use of the telescope, which magnified objects by a factor of 20. He was able to draw the moon's phases and showed that its surface was not smooth. The four main moons of Jupiter were observed, now referred to as the Galilean moons. He observed that the planet Venus goes through phases, the same as the moon. All these observations supported the view of Nicolaus Copernicus (1473–1543) that the earth was not the center of the universe and that the planets revolved around the sun. Galileo's observations were recorded in a book entitled *Siderius Nuncias* (the Starry Messenger). In the subsequent period, Galileo had to tread a fine line with the teachings of the church. He had been given permission to present his writings on the condition that he treat the Copernican theory as hypothetical. However, it was determined that he had breached this condition and that a case would be brought against him by the Inquisition. He was pronounced to be suspect of heresy, made to admit that he had overstated his case, and made to objure formally. Although he was condemned to life imprisonment, he moved to a villa in the hills above Florence where he lived fairly comfortably for the rest of his life.

## Dorothy Hodgkin (1910–1994)

Dorothy Hodgkin (née Crowfoot), of British nationality, was born in Cairo, Egypt, while her parents were working there. Her father was an archaeologist with the British Ministry of Education, and this gave Dorothy opportunities for travel in her early years. She graduated from Somerville College, Oxford, with a degree in chemistry. In 1933, she began work on her doctoral degree with the eminent scientist J.D. Bernal at the Cavendish Laboratory in Cambridge. Their field of study was the relatively new one of X-ray crystallography. Pioneering work by Max von Laue, William Henry Bragg, and William Lawrence Bragg had shown that atoms in a crystal deflected X-rays to give patterns that enabled information to be deduced about the positions of atoms in the crystal. Hodgkin made the technique of crystallography her own and used it to determine the structure of complex biological compounds including proteins. Her achievements in elucidating the structures of important molecules make an impressive record:

1937: Cholesterol
1945: Penicillin
1954: Vitamin $B_{12}$
1969: Insulin

In addition, she contributed to structure determinations of lacto-globulin, ferritin, and Tobacco Mosaic Virus (TMV). She was awarded the Nobel Prize for chemistry in 1964 for her research, especially on the structure of vitamin $B_{12}$. Her work on insulin demonstrated her ingenuity and relentless perseverance over a period of 34 years from 1935. Although the insulin molecule is small by protein standards, it is nevertheless a polymer of 51 amino acid residues, a daunting task for those who attempt to deduce its structure. Hodgkin's approach included replacement of the zinc atoms of insulin with atoms of three heavier elements, lead, uranium, and mercury, to facilitate the structural analysis. The relationship between the molecular structure of insulin and its specific function in regulating sugar metabolism has opened the way for understanding structure–function relations of complex biological molecules.

Hodgkin worked at Oxford University as well as in the crystallography group of Bernal at Cambridge. She hosted many students from overseas countries and formed lasting friendships with scientists abroad. She appears to have been perfectly at ease working in what was very much a male-dominated profession. The change in name upon marriage has at times been problematic for academic women. She published her penicillin studies under her maiden name, Crowfoot, and later announced the structure of vitamin $B_{12}$ as Hodgkin, her married name. Subsequently, some scientists did not know that they were the same. Many women avoid this problem by retaining their maiden name throughout their professional life. In addition to her scientific activities, she championed movements for world peace and disarmament.

Rather than further detailing her achievements, it may be more instructive to quote some of the comments made about her by contemporaries who knew her well. One of her colleagues, Max Perutz, who shared the Nobel Prize with John Kendrew for studies on the structure of globular proteins, said of her, "She will be remembered as a great chemist, a saintly, gentle and tolerant lover of people and a devoted protagonist of peace."* Another colleague (Dodson, 2002) had this to say about her: "Although Dorothy's laboratory was a centre for X-ray analysis and she was uncompromisingly dedicated to research, her warmth and kindness to the people in her laboratory made it an especially human place" (p. 181). One of her contemporaries commented that "Dorothy had an unerring instinct for sensing the most significant structural problems in the field, she had the audacity to attack these problems when they seemed well-nigh insoluble, she had the perseverance to struggle onward when others would have given up and she had the skill and imagination to solve these problems once the pieces of the puzzle began to take shape" (Glusker, 1994, p. 2465).

## Irving Langmuir (1881–1957)

Irving Langmuir gained a Bachelor of Science in metallurgical engineering from the Columbia University School of Mines, New York, in 1903. He had been greatly influenced by his older brother, Arthur, a research chemist, who had helped him set up his first chemical laboratory in the corner of his bedroom. In 1906, he obtained his Ph.D. under Nobel Laureate Walther Nernst in Gottingen. His Ph.D. research was done on the "Nernst glower," an electric lamp invented by Nernst. He then taught at Stevens Institute of Technology in Hoboken, New Jersey, until 1909, when he began work at the General Electric Research Laboratory in Schenectady, New York. One of the first problems that he worked on, under the direction of Willis Whitney, was the short life of the recently developed tungsten filament lamp, caused by a rapid and progressive blackening of the bulb. One of the lines of attack on the problem at the time was to try to improve the vacuum in the bulb. Langmuir convinced Whitney that he should tackle the problem by investigating the effects on the tungsten filament of the different gases that his colleagues were trying to eliminate. Conventional wisdom was that the blackening was caused by water vapor, and thus the need for low pressure. Langmuir bubbled gas through liquid air into the lamp and showed that the blackening was unaffected. Soon after, he discovered that the tungsten was being evaporated from the hot filament. Particles from the tungsten,

---

* San Diego Supercomputer Center, "Dorothy Crowfoot Hodgkin: A Founder of Protein Crystallography" (www.sdsc.edu/ScienceWomen/Hodgkin.html), accessed October 2010.

because of the high vacuum, could travel across the empty space and condense on the bulb, causing the blackening. The solution that Langmuir came up with was to surround the filament by a blanket of relatively inert gas. That would reduce the evaporation rate, making the filament last longer and enabling it to burn more brightly. At first, nitrogen gas was used, but later it was found that argon acted as the best blanket. The design of electric lamps subsequently survived relatively unchanged for a long period. Langmuir's approach had not been to tackle the problem head on. He had an inquiring mind that allowed him to consider all aspects of a phenomenon. Although he had kept the central problem at the back of his mind, he was motivated to study the fundamental theory of processes associated with it. These studies gave the understanding that was needed to arrive at a solution to the problem. They also led to applications in other areas, such as development of the high vacuum electron tube, the isolation of atomic hydrogen, and the process of hydrogen arc welding.

Subsequently, Langmuir's research extended into areas of science of great diversity, including valence theory, catalysis, surface films, and atmospheric physics. In 1932, he was awarded the Nobel Prize for his work on surface films. He was the first American industrial chemist to receive this honor. *The Collected Works of Irving Langmuir* (1962) is a collection of his publications in 12 volumes. The Langmuir Laboratory for Atmospheric Research near Socorro, New Mexico, was named in his honor. In surface chemistry, his name has been renowned by the Langmuir Adsorption Isotherm and by a journal of the American Chemical Society, called *Langmuir,* which is devoted to surface science. The legacy of Irving Langmuir is the brilliant science that resulted from his genius and the freedom he was given by the General Electric Company and his first director, Willis Whitney, to study the fundamental behavior of phenomena in which he was intensely interested.

*History proves abundantly that pure science, undertaken without regard to applications to human needs, is usually ultimately of direct benefit to mankind.*[*] (Irving Langmuir)

*The scientist is motivated primarily by curiosity and a desire for truth.*[†] (Irving Langmuir)

---

[*] Great-Quotes.com (www.great-quotes.com/quote/1440056), accessed September 2010.

[†] Great-Quotes.com (www.great-quotes.com/quote/1440059), accessed September 2010.

# Lise Meitner (1878–1968)

Lise Meitner was an Austrian physicist who was a member of the team that discovered nuclear fission. In her youth, women were not allowed to attend institutions of higher education. However, with support from her family, she was able to have private tuition that enabled her to attend university. She entered the University of Vienna in 1901 and, in 1905, completed her undergraduate studies and commenced her doctoral degree. She was influenced in her early scientific career by a famous scientist, Ludwig Boltzmann. After completion of her doctoral degree, encouraged by her father and with his financial assistance, she moved to Berlin. Here she was given the unusual privilege of attending the lectures of Max Planck. One year after her arrival in Berlin, she became Planck's assistant. In the following years, she began a lifelong collaboration and friendship with Otto Hahn, an outstanding chemist, and with him discovered several new isotopes. In 1909, she presented two papers on beta radiation (emission of electrons).

In 1912, the Hahn–Meitner team moved to the Kaiser Wilhelm Institute (KWI) in Berlin, which had recently been founded. Initially, Meitner worked without salary as a guest worker in Hahn's department of radiochemistry. In 1917, together with Hahn, she discovered the most abundant and longest-lived isotope of protactinium (Pa-231, half-life of 32,760 years). She was made head of the radio physics section at the KWI with Hahn as head of radiochemistry. In 1922, she discovered the cause of the effect now known as the Auger effect, after the French scientist, Pierre Auger, who observed the phenomenon in 1923. When an electron is removed from the core level of an atom to leave a vacancy, an electron from a higher energy level may fall into the vacancy resulting in a release of energy. The energy can be transferred to another electron (the Auger electron) that is ejected from the atom.

At that time, several groups were working on nuclear transitions, including Ernest Rutherford in Britain, Irene Joliot-Curie in France, and Enrico Fermi in Italy, in addition to the Meitner–Hahn team in Germany. In 1933, Adolph Hitler came to power, and soon after, persecution of Jews in Germany began. Meitner was from a Jewish family, although she and some of her siblings converted to Christianity later in life. Although many Jewish scientists left Germany, Meitner remained for some time. However, after the annexation of Austria, she escaped in

1938 through the Netherlands to Sweden. She took up a post at Manne Siegbohm's laboratory in Stockholm. She continued to correspond with Hahn, and the two had a clandestine meeting in Denmark. Hahn had found that neutron bombardment of uranium produced the element barium but, puzzled by this result, consulted with Meitner. Meitner, working with her nephew, Otto Frisch, was the first to articulate a theory of how the nucleus of an atom could be split into smaller parts. Uranium nuclei had split into barium and krypton with accompaniment of the ejection of several neutrons and a large amount of energy, the process being called nuclear fission. She realized that the source of the huge release of energy could be explained by Einstein's famous equation, $E = mc^2$, relating mass and energy interconversion, and that a chain reaction of explosive magnitude could be realized. As a consequence, several scientists convinced Albert Einstein to write a letter to U.S. President Franklin Roosevelt explaining the potential for a devastating weapon and warning of the danger of the knowledge falling into the wrong hands. Thus, the Manhattan Project at Los Alamos, New Mexico, was initiated with the purpose of producing an atomic bomb. Meitner refused to be involved in the project.

Meanwhile, Hahn and Fritz Strassman published the result that barium had been produced by bombarding uranium with neutrons in Naturwissenschaften in 1939. It was politically impossible for the exiled Meitner to publish jointly with Hahn. In 1944, Hahn received the Nobel Prize for chemistry for the discovery of nuclear fission. Many believed that Meitner should have shared the prize, but the circumstances prevented this. In 1966, Meitner shared the Enrico Fermi Award with Hahn and Strassman, going some of the way to compensating for being overlooked by the Nobel Committee. She was also recognized by receiving the award in 1946 of the "Woman of the Year" from the National Press Club (United States). She was awarded the Max Planck Medal of the German Physics Society in 1949 as well as honorary doctorates at Princeton University (New Jersey) and Harvard University (Cambridge, Massachusetts). Further recognition was given to her in 1997 when element 109 was named meitnerium in her honor.

## Gregor Mendel (1822–1884)

Gregor Mendel (born Johann Mendel) was born into a poor farming family in Hynice, Moravia (now Czech Republic). To escape a life of poverty, he entered the monastery of the Augustinian Order of St. Thomas where he took the name Gregor. He attended the University of Vienna to study teaching but failed his exams for a teaching diploma. His interest in the natural sciences led him to stay on at the monastery and use his time to carry out experimental work as well as teach mathematics. He chose the common garden pea for his research and grew these in the monastery garden. Peas had the advantage in that they could easily be grown in large numbers, and their reproduction could be manipulated. Mendel was able to selectively cross-pollinate purebred plants with particular traits and document the outcomes over many generations. Farmers had for thousands of years been breeding their plants in attempts to improve their quality, but it had been a hit-or-miss approach in which the mechanisms governing inheritance were not understood. Mendel's work addressed the question: "What really happens in the transmission of hereditary traits from parents to progeny?"

Mendel worked for several years, carefully self-pollinating and wrapping each individual plant to prevent accidental pollination by insects. Between 1856 and 1863, he cultivated and tested at least 28,000 pea plants, analyzing seven characteristics or traits of the pea plants. For example, he found that crossing tall and short parent plants gave progeny that resembled the tall parent, rather than being a medium height blend. This type of observation that traits do not show up in offspring plants with intermediate forms was critically important, because at the time, the accepted theory was that inherited traits blended from one generation to another. Mendel's work showed that this "blending theory" was wrong. In simplistic terms, his work showed that

1. The inheritance of each trait is determined by *units* or *factors* that are passed on to descendants unchanged. (These units are now called *genes*.)
2. An individual inherits one such unit for each trait.
3. A trait may not show up in an individual but can still be passed on to the next generation. This led to the concept that a dominant gene such as gives the tall pea plant will hide the recessive gene giving the short plant.

Mendel presented his work to the Natural History Society of Brno in a two-part paper in February and March 1865, and this was published in 1866 with the title "Experiments in Plant Hybridization."

The work was not understood and was largely ignored until the early 1900s when several scientists recognized its importance. However, it was not until the 1920s and 1930s that its full significance was appreciated. Since then, it has allowed scientists to predict the expression of traits on the basis of mathematical probability. It has become the foundation for the modern science of genetics. *Experiments in Plant Hybridization* is one of the most enduring and influential publications of science.

## Louis Pasteur (1822–1895)

Louis Pasteur was born in Dole, France, the son of a poor tanner. He gained his doctorate in 1847 at the École Normale Supérieure in Paris, where he started his research studying crystal structure, focusing on tartaric acid. He found that an aqueous solution of tartaric acid from living things, such as grape products, rotated the plane of polarized light. A solution of tartaric acid that had been chemically synthesized, on the other hand, did not show this effect, even though its elemental composition and chemical reactions were identical. Pasteur resolved this puzzle by separating the synthetic tartaric acid into two components (isomers), each of which rotated the plane of polarized light but in opposite directions. This property, called chirality, is caused by molecules that lack a plane of symmetry and have nonsuperimposable mirror images. Pasteur showed that, in general, where biological molecules have optical isomers, only one of the forms will be biologically active.

After several years teaching and carrying out research at Dijon and Strasbourg, he took the position of professor of chemistry at the University of Lille. Here he continued work on fermentation that he had started at Strasbourg. At that time, the consensus was that the process that caused spoilage of beverages such as beer, wine, and milk was caused by spontaneous creation. Pasteur exposed boiled broths to air in flasks that were designed to prevent particles from entering. Nothing grew unless the vessel was broken, showing that organisms (germs) that grew in the broths came from outside, supposedly as spores on dust. This debunked the widely accepted myth of spontaneous generation and gave strong support to the germ theory. Pasteur invented the process in which beverages were heated to kill bacteria and molds that were present, a process aptly called pasteurization.

This work led to the idea that microorganisms infected animals and humans to cause diseases. He urged the use of antiseptics to prevent their entry in surgery and turned his attention to disease prevention. This was influenced by the deaths of three of his five children from typhoid before reaching adulthood. During his work on chicken cholera, he instructed his assistant to inoculate some chickens while he was away on holidays. The assistant failed to carry out the instruction. On his return, the month-old cultures made the chickens unwell, but instead of being fatal, the chickens recovered completely. His assistant assumed that an error had been made and was about to discard the culture when Pasteur stopped him. He guessed that the recovered chickens might have become immune to the disease. Although it had been known that the weak form of a disease such as cowpox gave cross-immunity to the much more virulent smallpox, this was different. In this case, the weaker form of the chicken cholera had been generated artificially so that a naturally weak form of the disease did not need to be found. This revolutionized work on combating diseases and led to the development of vaccines. Pasteur went on to produce the first vaccine for rabies as well as chicken cholera and anthrax.

Pasteur's work served as the foundation for many branches of science and medicine, including stereochemistry, microbiology, bacteriology, virology, and immunology. It is doubtful if any other scientist has made such a hugely beneficial contribution to humankind.

*Imagination should give wings to our thoughts but we always need decisive experimental proof, and when the moment comes to draw conclusions and to interpret the gathered observations, imagination must be checked and documented by the factual results of the experiment.*[*]
(Louis Pasteur)

*In the field of observation, chance favors only the prepared mind.*[†]
(Louis Pasteur)

## Nikola Tesla (1856–1943)

Nikola Tesla was an inventor and engineer who was an ethnic Serb, born in what is now Croatia, later living in the United States

---

[*] Access Excellence (www.accessexcellence.org/AB/BC/Louis_Pasteur.html), accessed October 2010.

[†] Quote DB (www.quotedb.com/quotes/2195), accessed September 2010.

and taking American citizenship. He developed revolutionary advances in the field of electromagnetism in the late 19th and early 20th centuries. His large number of patents and theoretical work were the basis of modern alternating current (AC) electrical power systems, including the polyphase system of electrical distribution and the AC motor. He made other vital contributions that have influenced development of wireless communication, LASER, radar technology, X-rays, neon lighting, robotics, and remote control. The world today owes much to the genius of this man, but remarkably, relatively few people know his name even though his accomplishments were probably greater than those of Edison and Marconi, whose names are widely known.

Tesla studied at the Austrian Polytechnic in Graz and the Charles-Ferdinand University in Prague, but his academic achievements were not spectacular. However, he read many works, apparently being able to memorize complete books. As well as Serbian, he spoke Czech, English, French, German, Hungarian, Italian, and Latin. During his early life, he was stricken by several illnesses. He also suffered from a strange affliction in which he would see blinding flashes of light, sometimes accompanied by visions. The visions were linked by words or ideas he had come across at a previous time. This type of experience, now known as synesthia, occurs when stimulation of one sensory or cognitive pathway leads to automatic experiences in a second sensory or cognitive pathway. A simple example is when a particular written word may produce the vision of a specific color. It is an experience shared with others, and many report it as a gift that can aid in their creative processes. In his autobiography, Tesla describes an event where he was enjoying a walk with a friend in the City Park. It was at a time when his mind was focused on ways to generate electric current without the problems inherent in direct current (DC), such as sparking from brushes. He was reciting a passage from Goethe, which he found to be inspiring, when an idea came to him like a flash of lightning and the truth was revealed. He drew a diagram in the sand of the rotary magnetic field that was to be the basis of AC power generation. This same diagram was presented to the American Institute of Electric Engineers six years later. Tesla had the gift of visualizing his inventions in his mind. He had no need for drawings. In 1887, the Tesla Electric Company built motors exactly as he had visualized them. The pictures in his mind were exactly to scale and needed no attempts to improve the design.

In 1884, Tesla arrived in New York with a letter of recommendation to Thomas Edison. After working for a time, he parted company with Edison. Edison later admitted that his biggest mistake had been trying to develop direct current, rather than the superior alternating current system that Tesla had put within his grasp. Tesla worked for a time as a laborer, digging ditches and raising capital for his next project. In 1887, he constructed the initial brushless alternating current induction motor. He joined with George Westinghouse at Westinghouse Electric and Manufacturing Company's Pittsburgh, Pennsylvania, laboratories. Here he was able to develop his ideas for polyphase systems, which allow transmission of AC electricity over long distances.

During his career, Tesla had many setbacks. These included serious illnesses and a fire in 1898 that destroyed records and put back his work on wireless radio transmission which otherwise may have given him priority over Marconi. He died penniless in his apartment in New York and without the fame that he so richly deserved.

*Every effort under compulsion demands sacrifice of life energy. I never paid such a price. I have thrived on my thoughts.** (Nikola Tesla)

# References

Darwin, C. 1859. "On the Origin of Species by Means of Natural Selection, Or the Preservation of Favoured Races in the Struggle for Life" (www.gutenberg.org/books/1228), accessed October 2010.

Dodson, G. 2002. "Dorothy Mary Crowfoot Hodgkin O.M. 12 May 20 1910–29 July 1994." *Biographical Memoirs of Fellows of the Royal Society* 48:179–219.

Glusker, J.P. 1994. "Dorothy Crowfoot Hodgkin." *Protein Science* 3:2465–2469.

Langmuir, I. 1962. *The Collected Works of Irving Langmuir*, Pergamon Press, Elmsford, New York.

Lyell, C. 1830–1833. *Principles of Geology, Volumes 1–3* (www.esp.org/books/lyell/Principles/facsimile/title3.html), accessed October 2010.

Malthus, 1798. "An Essay on the Principle of Population" (www.econlib.org/library/Malthus/malplong.html), accessed October 2010.

Mendel, G. 1866. *Experiments in Plant Hybridization* (www.mendelweb.org/Mendel.html), accessed October 2010.

---

* Tesla Universe (www.teslauniverse.com/nikola-tesla-quote-42), accessed October 2010.

# CHAPTER 8

# Future Challenges for Scientific Research

I will not cease from mental fight, nor shall my sword sleep in my hand.*

**William Blake**

What are the main challenges facing scientific research in the immediate future? Until now, we have been considering topics such as the training that a scientist needs to undergo for a successful career, the attributes that need to be developed, and the way the scientific method should be applied. The focus has been on the individual. In this chapter, I would like to focus more generally on what is needed for scientific research to continue to flourish and contribute positively to human progress. Of course, the way science evolves is dependent on the cumulative efforts of individuals, both active scientists and the nonscientists who influence policies.

## Two Areas for Change in Direction

There are two main areas where I believe there is a need to change the direction in which scientific research is heading. First, it is desirable for the shackles that have been imposed on the creativity of scientists in recent times to be loosened. Second, there is a need to nurture a more pleasant environment for scientists to work in than the current one that is experienced in many scientific institutions. The two problems mentioned are not mutually exclusive. In fact, they are interrelated and stem from the same source. This source is the controls that have been placed on scientists in the past few decades resulting from the managerial ideology discussed in Chapter 5.

---

* GoodReads.com (www.goodreads.com/quotes/show/45356), accessed October 2010.

Some readers may think that this is just the opinion of someone who is pushing for a better deal for scientists. After all, why should scientists not be controlled? If they are not, they may well spend their time in ivory towers following their whims and not producing anything of value to humankind. This can happen. Other readers may question why a book that is supposedly aimed mainly at giving advice to budding scientists should be concerned with science policy, which is more the domain of politicians and bureaucrats. I will attempt to answer these questions in this chapter. Briefly, the curtailing of freedom for scientists to develop their creativity not only stifles scientific progress but, as a consequence, erodes future economic growth. Every working scientist and all those contemplating a career in science should be aware of this threat. As mentioned, the direction of scientific research in the future will be determined by the cumulative input of all those who have a role to play. It will not be sufficient for researchers beginning their careers to focus only on perfecting their scientific skills. They will also need to take a principled position in regard to how the ideals of science should be upheld. The future direction of science will depend on whether they stand up to the challenge or whether their response is pusillanimous.

## Why Are Humans the Only Species to Have Progressed Culturally?

In his book *Pioneering Research*, Donald Braben (2004) presents what I believe should be compulsory reading for all those embarking on a research career. Here I can only summarize briefly the main ideas of the book. Braben traces the evolution of humankind from its primitive beginnings to the present. Humanoids were physically inferior to some other animals in many respects, yet they have managed not only to survive but to become the dominant species of the planet. While other species have remained the same for millions of years, humans have undergone a spectacular transformation. What is the explanation? Braben suggests that the trait responsible for the unique progress of humankind is an indomitable spirit of dissent possessed by some of its members (that is, an innate tendency not to accept how things are). This has been the driving force for the changes that have propelled civilization. These changes have come primarily through science and technology. Of course, this is not to play down the role that other human endeavors, such as art and literature, have contributed to civilization. However, without the technological advances that have been made, as Braben points out, humans would still be trying to survive in a tooth-and-nail struggle. The most significant advances in science have come from those who have had the freedom to pursue their inquiries. Great discoveries have usually come out of the blue. Here, it is necessary not to confuse what is meant by *out of the blue*. It does not refer to someone wandering aimlessly and suddenly stumbling on an important finding. The important findings are discovered by researchers who have acquired advanced knowledge of a discipline and are thoroughly conversant with what has been found by previous workers. Then, armed with this knowledge, they relentlessly pursue ideas wherever they lead until insights into previously unimagined territory are revealed.

## Why Present Funding Procedures for Research Are Unsatisfactory

Braben submits that present conditions for scientists largely eliminate this type of research. In the past several decades, as a result of the increasing emphasis placed on efficiency and accountability, the creativity of researchers has been greatly curtailed. This is exemplified in the way scientific research is currently funded. There are now many more scientists competing for a finite amount of funds. Funding bodies require scientists to submit applications for funding that must conform to strict guidelines. These are then evaluated by peer review. The fallacy in this process has been pointed out in Chapter 5, referring to comments of Albert Szent-Gyorgi (1974). It tends to promote pedestrian research and does not allow for the unexpected discoveries that lead to the most important scientific breakthroughs. Research proposals submitted to funding bodies are usually required to set out their aims and to include descriptions of expected outcomes, milestones, and time lines for completion of the different stages.

The adoption of peer-review processes for selection of proposals to fund has been in operation for a relatively short time, several decades. It has resulted in a burgeoning bureaucracy. Scientists need to spend a large amount of time in preparing their proposals, and because only a small proportion is funded, this results in a huge waste of time for the majority who are unsuccessful. To be approved, the proposals must convince peer reviewers that the work is feasible and that the objectives can be achieved. In many cases, much of the experimental work needs to be completed to demonstrate to reviewers that it can be viable. Thus, we often have the absurd situation where scientists have to practically complete the work to convince peer reviewers that it is viable, so that funding will be provided to do work that has already been done.

The shackles that have been imposed on creative research as a result of peer reviewing of proposals must surely induce a mind-set for scientists to plan unambitious objectives that can be accomplished without too much risk. This means that no room is left for following up the free thinking that, in previous times, has led to the most important advances in knowledge. The negative effects on scientific progress have been well described by Braben (2004), who has suggested an alternative to the peer review system and has put this into practice with the Venture Research Program (Braben, 2004, 2008). Initially funded by B.P., research projects are chosen for funding support on the basis that they are outside the mainstream of the proposals submitted for peer reviewing.

In my own field of research, there has been a decrease in the number of scientists who use a fundamental approach. These are the ones who have contributed basic new knowledge. As they retire, they are not being replaced by scientists with a similar fundamental approach. As a result of the increasing need for university departments and scientific organizations to acquire their own research funding, the selection criteria for new appointees are giving a heavy weighting to success in obtaining funding. This attribute is also reflected in career advancement. A scientist who obtains a number of substantial grants may expect to be promoted rapidly to higher levels. The research may be fairly pedestrian, encompassing, as we have seen, projects with simple objectives that may be safely achieved. The result is an increase in short-term mundane research and disappearance of the research needed to provide the foundation for significant advances. Inevitably, this is a path to mediocrity. How this trend can be reversed is a challenge for future leaders in science.

In Chapter 7, we looked briefly at the careers of some notable scientists whose contributions to science have been outstanding.

Most of these men and women, whose efforts have revolutionized human progress, would not have been able to make their contributions if they had to participate in today's competition for the funding required to do their research. Before he set out on his unplanned quest, Albert Einstein did not know what he would discover, yet his achievements were awesome. Similarly, Charles Darwin began his observations without any preconceived notions of where they would lead him. Both these scientists, whose thinking profoundly affected the human world, had the freedom to pursue their innate curiosity. Irving Langmuir, an industrial scientist, was given free rein for his genius to flourish. No scientist working in industry today could hope for this freedom and therefore could not hope to match the accomplishments of Langmuir. Dorothy Hodgkin began working on the structure of insulin in 1935 and completed this daunting task in 1969. If she had submitted a proposal to elucidate the structure of insulin over that period, would she have satisfied the peer-reviewing process? At the start of the work, the obstacles to be overcome could not be conceived, so it would have been absurd to try to propose expected outcomes and time lines for achieving objectives. All the scientists mentioned in Chapter 7 could be thought of as dissenters according to the views of Braben (2004). They did not accept the *status quo* of knowledge at that time and reached out to extend the boundaries of our understanding. They could only do this because of the intellectual freedom they enjoyed that enabled them to explore the unknown.

## Stifling of Creativity in Science Can Stunt Future Economic Growth

Earlier in the chapter, I alluded to the effect that stifling of creativity in science could have in stunting economic growth. How could this be? Braben explains very well how this may happen. The world's population is increasing, and there is not a lot that can be done to halt or even slow the rate of increase. This puts increasing pressure on resources, particularly food supplies, needed to at least maintain the standard of living. In the past, this has been achieved by corresponding increases in economic advances. Economic growth is a complex parameter affected by many variables, most of which are difficult to quantify or to predict. Most measures that have been used, such as real global gross domestic product (GDP), show a stochastic behavior over time. Economists have attempted to explain the complex question of how economic growth happens. Some 50 years ago, Robert Solow (1957) showed how new technologies created a large portion (estimated to be about four fifths) of economic growth. For this work, he was awarded the Nobel Prize in economics in 1987. Not all growth economists agree about the fraction of growth that can be explained by technological progress, but all do agree that its contribution is important. We need to remember that most technology stems from innovative basic science.

Braben (2004) has concluded from analysis of data that world GDP per capita was constant or increasing from the 1950s to the early 1970s. This period is often referred to as the Golden Age, because global economic growth was so high. This period also coincided with scientific and technological advances—semiconductors, lasers, nuclear power generation, computers, and plastics, to name just a few. Other spectacular events such as sending satellites into orbit and traveling to the moon happened in that era. Braben suggests that since the end of the Golden Age in the early 1970s, there has been a decline in economic growth as quantified by the world real GDP per capita. This period of decline correlates with the time since scientific research has been subjected to increasing control. Researchers no longer have the freedom to follow their ideas and explore the unknown, but instead have had to conform to strictly regulated programs. The protocols used for selection and support of research projects inhibit adventurous studies that involve major departures from the beaten track.

## Suppression of Freedom Causes Stagnation of Knowledge

We can also learn from history. Some isolated societies that we would regard as primitive do not show substantial changes over long periods of time. Their customs and folklore are handed down from one generation to the next and are accepted without question. For societies that we would consider more advanced, periods where there has been economic and cultural stagnation have corresponded to times of suppression of freedom and knowledge. These periods have been imposed by authoritarian rule, be it of political or religious origin. The Dark Ages of Europe in the time from the decline of the Roman Empire (around AD 450) to about AD 1000 or longer were characterized by submission to dogma and elimination of inquiry. This changed in the 17th and 18th centuries (Age of Enlightenment) when people rebelled intellectually and began to question everything. From the 8th to the 14th centuries, Muslim scientists and scholars made great contributions to human knowledge. The splendid Muslim art and architecture of this period can still be appreciated in Andalucia in Spain. Why has Muslim science and culture not progressed from those times as it has in Western countries?

## The Need for a Change in the Working Environment for Research

The other area where I feel there is a need for change is the working environment that many scientists find themselves subjected to. This mainly stems from the command-and-control climate that has been foisted on them. Scientists are increasingly prevented from exercising the freedom that allows them to be creative. They are treated as "knowledge workers" whose contributions are measured by how well they comply with objectives that they are expected to complete in a certain time. Their productivity is assessed by similar measures that are applied to those who work on a factory production line. Goals that are achieved in a short time are those that are favored as making a quick contribution to the bottom line. For a scientist to work under these conditions can be a source of great frustration and results in lowering of morale and often to health problems as was described in Chapter 5.

The hierarchical command-and-control culture has encroached on scientific research institutions hand-in-hand with the managerial organization of research. Unfortunately, those who aspire to positions of control are often motivated by greed and the desire for power. These types of people have little empathy for subordinates. Rather than supporting and encouraging them, they may trample on their best interests if they feel that they pose a threat to their authority. How can this trend toward an unpleasant workplace environment be turned around? There is no easy answer, but this does not mean that one should not try. In his book *Corporate Renaissance*, Rolf Osterberg (1993) writes about the "Old Thought" and the "New Thought" in business. The "Old Thought" is what we have been experiencing, and the "New Thought" is a shift in thinking toward valuing the individual and creating an environment that is wholesome and compassionate. This is what he has to say about the hierarchical system: "The hierarchical system is a power structure. It is built upon fear, suspicion and lack of trust. It reinforces the distance among individuals and thoroughly prohibits creative cooperation by creating a climate of competition. It is an impediment to development and leads to wasted energy. It builds inertia and inflexibility into our organizations and is devastating to our creativity. We will never realize our full potential as human beings unless hierarchies are abandoned" (Osterberg, 1993, p. 48). If this has been said about the business world, how much more pertinent is it to be said of the scientific world?

For those entering a scientific research career, how will they respond to the challenge I have posed? Will some dismiss it as being overstated or irrelevant? Perhaps some will be excited by the prospect of using the system to attain the power to subjugate others. Will some react with indifference and, if they admit there is a problem, leave the problem for others to deal with? Or will there be some who recognize the problems and are ready to accept the challenge and be courageous enough to try to change things for the better? The latter option is the one that may well be the most difficult, but it may also be the most noble.

Improvement in the way scientific research is organized and in the working environment for scientists will only occur if the managerial system is made more flexible. As long as those who hold positions of authority in scientific organizations believe that the way they operate is correct, conditions will not change and will probably worsen. For a managerial ideology to have taken control of scientific establishments, certain conclusions can be drawn. Scientists who have allowed it to happen must not have a good understanding of how the scientific process works, or they have been too pusillanimous to resist the changes. If this is the case, how easy will it be, now that managerialism is well entrenched, to effect any change?

Nevertheless, those who recognize how science is being misdirected are duty bound to try to turn things around. Scientists need to engage more with those who hold positions of authority, those who control the finances, and the politicians and members of the public who influence policy. The manner in which scientific research should be conducted is rarely debated publically. Many outside the scientific community may not be aware of the dangers posed to discovery of new knowledge by currently imposed controls. To those who have never thought very much about research, it makes sense to manage it so as to have clear objectives and to document progress in terms of meeting the objectives within certain time frames and predicting the financial benefits that will accrue. They have little understanding of how this approach leads to nothing more than achieving what was planned initially. In contrast, the spectacular breakthroughs that have had the greatest impact on human progress have resulted from imaginative thinking that has not been constrained by preconceived objectives. Elimination of the latter type of research, which is what is currently happening, is, to use the words of Donald Braben, killing the geese that lay the golden eggs. Those people who understand this have the obligation to explain it and to argue the case for changes in policy.

Changes of policy that lead to less regimentation of research scientists and greater opportunity for them to realize their true potential are bound to benefit the general community. But this will not be achieved easily. Those scientists who support such changes have work to do to bring the issues into the public domain and to have them debated. Any small movement back toward more freedom for scientists will inevitably be accompanied by a healthier workplace environment. The workplace of research institutions would then be dominated less by autocratic managers and more by servant leaders with genuine concern for the welfare of their colleagues.

# References

Braben, D.W. 2004. *Pioneering Research. A Risk Worth Taking*, John Wiley & Sons, Hoboken, New Jersey.

Braben, D.W. 2008. *Scientific Freedom: The Elixir of Civilization*, John Wiley & Sons, Hoboken, New Jersey.

Osterberg, R. 1993. *Corporate Renaissance*, Nataraj, Mill Valley, California.

Solow, R.M. 1957. "Technical Change and the Aggregate Production Function." *The Review of Economics and Statistics* 39(3):312–320.

Szent-Gyorgi, A. 1974. "Research Grants." *Perspectives in Biology and Medicine* 18:41–43.

# Index

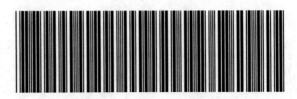